NOTICE.

A copy of this pamphlet will be furnished to each County Clerk, Town Clerk, Clerk of the Board of Supervisors, and Commissioners of Highways in the several towns in the State, in their official capacity. It will therefore belong to THEIR RESPECTIVE OFFICES, AND MUST BE HANDED OVER TO THEIR SUCCESSORS.

LAWS

OF THE

STATE OF NEW YORK

RELATING TO

HIGHWAYS, BRIDGES AND FERRIES,

COMPRISING

CHAPTER SIXTEEN OF THE FIRST PART OF THE THIRD OR REVISERS' EDITION OF THE REVISED STATUTES, WITH THE SEVERAL ACTS WHICH HAVE BEEN SUBSEQUENTLY PASSED ON THE SAME SUBJECT, WITH

FORMS AND INSTRUCTIONS;

ALSO

GENERAL LAWS

RELATING TO

PLANK AND TURNPIKE ROADS.

PREPARED UNDER THE DIRECTION OF THE

SECRETARY OF STATE,

PURSUANT TO A CONCURRENT RESOLUTION OF THE LEGISLATURE.

ALBANY:
WEED, PARSONS & COMPANY, PRINTERS.
1860.

COMPILER'S NOTE.

The Compiler has adopted the Revisers' plan of numbering anew the sections, retaining, however, within brackets, the original numbers; and for the purpose of enabling the reader at a glance to determine whether the text is part of the Revised Statutes, or of an act subsequently passed, the ordinary section mark [§] is placed before each original section, while a section mark preceded by an asterisk [*§] is placed before each section of laws passed since the revision. In cases where a whole title or article has been introduced, the numbers of the original sections, if changed, are shown within brackets, or by notes at the ends of sections.

CONCURRENT RESOLUTION.

Resolved (if the senate concur), That the secretary of state be directed to cause all of the laws of this state, relating to highways, bridges and ferries, to be collected and published in pamphlet form, with such notes and explanations as in his opinion are necessary for the understanding of the same; and that he cause five thousand copies of the same to be printed, of which he shall distribute one copy to each county clerk, town clerk, clerk of the board of supervisors and commissioner of highways, through the clerks of the several counties; and the remaining copies shall be held by him subject to the direction of the legislature.

STATE OF NEW YORK,
IN ASSEMBLY, *March* 3, 1860.

The foregoing resolution was duly passed.

By order,

WM. RICHARDSON, *Clerk*.

STATE OF NEW YORK,
IN SENATE, *April* 5, 1860.

The foregoing resolution was duly passed.

By order,

JAS. TERWILLIGER, *Clerk*.

CHAPTER XVI.

OF HIGHWAYS, BRIDGES AND FERRIES.

[Passed 3d December, 1827, and took effect 1st January, 1828.]

TITLE I.

OF HIGHWAYS AND BRIDGES.

ARTICLE I.

OF THE OFFICERS ENTRUSTED WITH THE CARE AND SUPERINTENDENCE OF HIGHWAYS AND BRIDGES, AND THEIR GENERAL POWERS AND DUTIES.

Section 1. The commissioners of highways in the several towns in this state, shall have the care and superintendence of the highways and bridges therein; and it shall be their duty: Commissioners, their duty.

2 Hill, 467; 6 do 463. 1. To give directions for the repairing of the roads and bridges, within their respective towns;

2. To regulate the roads already laid out, and to alter such of them as they, or a majority of them, shall deem inconvenient;

24 Wend., 491. 3. To cause such of the roads used as highways, as shall have been laid out but not sufficiently described, and such as shall have been used for twenty years but not recorded, to be ascertained, described and entered of record in the town clerk's office;

7 Wend., 476. 4. To cause the highways, and the bridges which are or may be erected over streams intersecting highways, to be kept in repair;

4 Hill, 593. 5. To divide their respective towns into so many road districts as they shall judge convenient, by writing under their hands, to be lodged with the town clerk, and by him to be entered in the town book; such division to be made annually, if they shall think it necessary, and in all cases to be made at least ten days before the annual town meeting;

Highway work. * 6. To assign to each of the said road districts such of the inhabitants liable to work on highways as they shall think proper, having regard to proximity of residence as much as may be; provided, however, that whenever the commissioners of any town shall have neglected, for the period of one year at any time after any public road or highway shall have been laid out and title thereto acquired by due process of law, to open or work the same or any part thereof, and whenever any number of inhabitants of any town, in or through which the said road has been laid out, shall have given ten days' notice to the commissioners of said town that they desire to apply the whole or any part of their highway labor to the working of said road, the said commissioners shall forthwith assign the said inhabitants to such road, direct the highway labor, for which they are annually assessed,

* As amended by chap. 63, Laws of 1853.

to be applied to the same, and cause the same to be worked and put in good order for vehicles and travelers, within one year, under the direction of any of the said inhabitants whom such commissioners may appoint as an overseer of the labor so to be applied to such road; and when the number of days' labor assessed in the current year to such inhabitants, as their annual highway tax, is not sufficient to put such road in good order, as aforesaid, then the said inhabitants may anticipate the whole or any part of the highway labor assessed, and to be assessed against them, for a period not exceeding three years; but from no one of the districts into which the said town is divided shall more than one-half of its annual labor be taken and applied to any road not embraced in said district;

7. To require the overseers of highways, from time to time, and as often as they shall deem necessary, to warn all persons assessed to work on highways, to come and work thereon, with such implements, carriages, cattle or sleds as the said commissioners, or any one of them, shall direct.

To lay out and discontinue roads.

§ 2. The commissioners of highways shall have power, in the manner and under the restrictions hereinafter provided, to lay out, on actual survey, such new roads in their respective towns, as they may deem necessary and proper; and to discontinue such old roads and highways, as shall appear to them, on the oaths of twelve freeholders of the same town, to have become unnecessary.

To account.

§ 3. The commissioners of highways of each town shall render to the board of town auditors, at their annual meeting for auditing the accounts of town officers, an account in writing, stating:

1. The labor assessed and performed in such town;

2. The sums received by such commissioners for fines and commutations, and all other moneys received under this chapter;

3. The improvements which have been made on the roads and bridges in their town, during the year imme-

diately preceding such report, and an account of the state of such roads and bridges; and,

4. A statement of the improvements necessary to be made on such roads and bridges, and an estimate of the probable expense of making such improvements, beyond what the labor to be assessed in that year will accomplish.

Repairs of roads and bridges.

§ 4. The commissioners of highways of each town shall deliver to the supervisor of each town, a statement of the improvements necessary to be made on the roads and bridges, together with the probable expense thereof; which supervisor shall lay the same before the board of supervisors at their next meeting. The board of supervisors shall cause the amount so estimated, to be assessed, levied and collected, in such town, in the same manner as other town charges; but the moneys to be raised in any such town, shall not exceed, in any one year, the sum of two hundred and fifty dollars.

1 Hill, 50.

Mile stones.

§ 5. It shall be the duty of the commissioners of highways of each town, to cause mile-boards or stones to be erected, where not already erected, on the post roads, and such other public roads in their town as they may think proper, at the distance of one mile from each other, with such fair and legible inscriptions as they may think proper.

Overseers, their duty.

§ 6. It shall be the duty of the overseers of highways in each town:

11 Wend., 667.

1. To repair and keep in order the highways within the several districts for which they shall have been elected;

2. When so required by the commissioners of highways, or any one of them, to warn all persons assessed to work on the highways in their respective districts, to come and work thereon;

3. To cause the noxious weeds on each side of the highway within their respective districts, to be cut down or destroyed twice in each year, once before the

first day of July, and again before the first day of September; and the requisite labor shall be considered highway work; and,

4. To collect all fines and commutation money, and to execute all lawful orders of the commissioners.

Overseers, their duty.

§ 7. It shall be the further duty of the overseers of highways, once in every month, from the first day of April until the first day of December, to cause all the loose stones lying on the beaten track of every road within their respective districts, to be removed; and to cause the monuments erected or to be erected as the boundaries of highways, to be kept up and renewed, so that the extent of such roads may be publicly known.

New assessments to be made by them.

§ 8. When the quantity of labor assessed on the inhabitants of any road district by the commissioners, shall be deemed insufficient by the overseer of such district to keep the roads therein in repair, it shall be the further duty of such overseer to make another assessment on the actual residents in such district, in the same proportion, as near as may be, and not exceeding one-third of the number of days assessed in the same year by the commissioners on the inhabitants of such district; and the labor so assessed by an overseer, shall be performed or commuted for, in like manner as if the same had been assessed by the commissioners of highways.

Guide-posts

§ 9. The commissioners of highways of each town shall cause guide-posts, with proper inscriptions and devices, to be erected at the intersections of all the post-roads in their town, and at the intersection of such other roads therein as they may deem necessary.

Id.

§ 10. It shall be the duty of the overseers of highways of each town to maintain and keep in repair, at the expense of the town, such guide-posts as may have been erected by order of the commissioners, within the limits of the districts for which they shall have been respectively elected or appointed.

Scrapers & ploughs.

§ 11. The commissioners of highways, whenever they shall think it necessary or useful, may direct and empower any overseer of highways, in their respective towns, to procure a good and sufficient iron or steel-shod scraper, and plough, or either of them, for the use of his road district; to be paid for, by the moneys arising from commutations and fines within such district.

Id.

§ 12. In case such moneys shall be insufficient for the purpose, the deficiency shall be assessed by the overseers upon the inhabitants of the districts, in the proportion they are respectively assessed on the assessment roll of said town; and if any one so assessed, shall neglect or refuse to pay such assessment, the same may be sued for and recovered by the overseer.

Excess of work by overseers.

§ 13. If any overseer shall be employed more days in executing the several duties enjoined on him by this chapter, than he is assessed to work on the highway, he shall be paid for the excess at the rate of seventy-five cents per day, and be allowed to retain the same out of the moneys which may come into his hands for fines under this chapter; but he shall not be permitted to commute for the days he is assessed.

Vacancy in their office.

§ 14. If any person chosen to the office of overseer of highways, shall refuse to serve, or if his office shall become vacant, the commissioners of highways of the town shall, by warrant under their hands, appoint some other person in his stead; and the overseer so appointed, shall have the same powers, be subject to the same orders, and liable to the same penalties, as overseers chosen in town meetings.

Proceedings.

§ 15. The commissioners making the appointment, shall cause such warrant to be forthwith filed in the office of the town clerk, who shall give notice to the person appointed as in other cases.

Penalties on overseers.

§ 16. Every overseer of highways who shall refuse or neglect, either:

1. To warn the people assessed to work on the high-

ways, when he shall have been required so to do by the commissioners, or either of them;

2. To collect the moneys that may arise from fines or commutations; or,

3. To perform any of the duties required by this chapter, or which may be enjoined on him by the commissioners of highways of his town, and for the omission of which a penalty is not hereinafter provided:

Shall, for every such refusal or neglect, forfeit the sum of ten dollars, to be sued for by the commissioners of highways of the town; and when recovered, to be applied by them in making and improving the roads and bridges therein.

§ 17. It shall be the duty of the commissioners of highways of each town, whenever any person resident in their town shall make complaint that any overseer of highways in such town has refused or neglected to perform any of the duties enumerated in the last preceding section, and shall give or offer to such commissioners sufficient security to indemnify them against the costs which may be incurred in prosecuting for the penalty annexed to such refusal or neglect, forthwith to prosecute such overseer for the offense complained of.

To be prosecuted by commissioners.

§ 18. If such commissioners of highways shall refuse or neglect to prosecute for such penalty, they shall, in every such case, forfeit the sum of ten dollars, to be recovered by the person who shall have made such complaint, and given or offered such security.

Penalty for neglect.

11 Wend., 867.

ARTICLE II.

OF THE PERSONS LIABLE TO WORK ON HIGHWAYS, AND THE MAKING OF ASSESSMENTS THEREFOR.

§ 19. Every person owning or occupying land in the town in which he or she resides, and every male inhabitant above the age of twenty-one years residing in the town, when the assessment is made, shall be assessed to

Persons liable to be assessed.

12 Wend., 390.

work on the public highways in such town; and the lands of non-residents, situated in such town, shall be assessed for highway labor, as hereinafter directed.

Meetings of commissioners.

§ 20. The commissioners of highways of each town shall meet, within eighteen days after they shall be chosen, at the place of town meeting, on such day as they shall agree upon, and afterwards at such other times and places as they shall think proper.

Lists of inhabitants.

§ 21. Each of the overseers of highways shall deliver to the clerk of the town, within sixteen days after his election or appointment, a list subscribed by such overseer, of the names of all the inhabitants in his road district, who are liable to work on the highways.

Non-resident lands, how appraised.

§ 22. The commissioners of highways in each town, at their first or any subsequent meeting, shall make out a list and statement of the contents of all lots, pieces or parcels of land within such town, owned by non-residents therein; every lot so designated shall be described in the same manner as is required from assessors, and its value shall be set down opposite to such description; such value shall be the same as was affixed to such lot in the last assessment roll of the town; and if such lot was not separately valued in such roll, then in proportion to the valuation which shall have been affixed to the whole tract of which such lot shall be a part. [*Sec.* 2 *of chap.* 154 *of* 1835.]

Lists of inhabitants.

§ 23. The town clerk shall deliver the lists filed by the overseers, to the commissioners of highways of the town; who shall proceed, at their next meeting, or at some subsequent meeting, to ascertain, estimate and assess the highway labor to be performed in their town the then ensuing year.

Proceedings in making assessments.

§ 24. In making such estimate and assessment, the commissioners shall proceed as follows:

1. The whole number of days' work to be assessed in each year shall be ascertained, and shall be at least three times the number of taxable inhabitants in such town;

2. Every male inhabitant being above the age of twenty-one years (excepting ministers of the gospel and priests of every denomination, paupers, and idiots and lunatics), shall be assessed at least one day;

3. The residue of such days' work shall be apportioned upon the estate, real and personal, of every inhabitant of such town, as the same shall appear by the last assessment roll of the said town; and upon each tract or parcel of land, of which the owners are non-residents, contained in the list made as aforesaid;

4. If, after such apportionment, there shall be any deficiency in the number of days' work determined by the commissioners to be performed in their town the then ensuing year, such deficiency shall be assessed upon the estates, real and personal, of the inhabitants of the town, and upon each tract or parcel of land of which the owners are non-residents, according to the last assessment roll;

5. The commissioners shall affix to the name of each person named in the list furnished by the overseers, and also to the description of each tract or parcel of land contained in the list prepared by them of non-resident lands, the number of days which such person or tract shall be assessed for highway labor, as herein directed; and the commissioners shall subscribe such lists, and file them with the town clerk. [*Sec.* 3 *of chap.* 154 *of* 1835.]

Stock or moneyed corporations to be assessed.

§ 25. In making the estimate and assessment of the residue of the highway labor to be performed in their town, after assessing at least one day's work upon each of the male inhabitants therein, above the age of twenty-one years, as provided in the sixteenth chapter of the first part of the Revised Statutes, entitled "Of highways and bridges," the commissioners of highways shall include among the inhabitants of such town, among whom such residue is to be apportioned, all moneyed or stock corporations, which shall appear, on the last assessment roll of their town, to have been assessed therein. [*Sec.* 1 *of chap.* 431 *of* 1837.]

Notice to be given to corporations.

§ 26. Such corporations shall be notified to furnish the amount of highway labor assessed to them in the same manner as individuals residing in such town, by giving oral or written notice to the president, cashier, agents, treasurer, or secretary of such corporation, or any clerk or other officer thereof, at the principal office or place of transacting the business or concerns of the said company; which labor shall be performed in such district or districts as the commissioners of highways of the town shall direct, and any number of days' work, not exceeding fifty, may be required to be performed by any such corporation in any one day. [*Sec.* 2, *of same chapter.*]

Corporations may commute.

§ 27. Every such corparation may commute for the highway labor assessed upon it, in the same manner and at the same rate as is allowed by law to individuals, or by paying such commutation to a commissioner of highways of the town; and the commutation money so paid, may be expended by the commissioners of highways upon any district or districts in the town; and for that purpose the said commissioners shall be entitled to demand and receive from the overseers, to whom any such commutation may have been paid, the whole or any portion thereof; but in every case where any such corporation shall be located in any city, village or town, where by law the road tax is now payable in money, the road tax imposed on any such corporation shall be paid in money, according to the provisions of the several laws affecting said city, village or town. [*Sec.* 3, *of same chapter.*]

Penalties for default.

§ 28. Such corporations shall be liable to the same penalties for every day's work required, and for every default of any substitute sent by them, as is provided by law in the case of individuals required to work on highways; which shall be collected in the same manner, and paid over to the commissioners of highways of the town, by the constable collecting the same, and may be expended by them in the same manner as herein provided for the commutation money received from any such corporation. The summons issued by any justice according to this act, may be for any number of penalties incurred by any such corporation previous thereto, and may be served in the manner provided by law for the service of writs or summons issuing out of courts of record against corporations. [*Sec.* 4, *of same chapter.*]

How to be collected.

§ 29. In case any such penalty cannot be collected as herein provided, the commissioners of highways of the town may file a bill in the court of chancery, against any such delinquent corporation, for the discovery and sequestration of its property; whereupon the same proceedings shall be had as are provided by law for the collection of county taxes assessed against incorporated companies, and the chancellor shall possess the like powers in respect to the same; and the said commissioners may also recover such penalties, or any number of them that may have been incurred, with costs, from such delinquent company, in any court of record in this state. [*Sec.* 5, *of same chapter.*]

Omissions, how to be rectified.

§ 30. Whenever the assessors of any town shall have omitted to assess any inhabitant or property in such town, the commissioners of highways shall assess the persons and property so omitted, and shall apportion highway labor upon such persons or property in the same manner as if they had been duly assessed upon the last assessment roll. [*Sec.* 6, *of same chapter.*]

Powers of commissioners of highways vested in certain other officers.

§ 31. When by law the powers and duties of commissioners of highways are conferred and imposed upon other officers, they shall possess all the powers and perform all the duties in this act conferred or enjoined upon such commissioners; and the assessments under it for the present year shall be made before the first day of July next, by the commissioners of highways. [*Sec.* 7, *of same chapter.*]

Non-residents.

§ 32. Lands of non-residents within any town, occupied and improved by the owner or owners, or his or their servants or agents, shall be liable to the same assessments for highways as if the owner or owners were residents. [*Sec.* 1 *of chap.* 107 *of* 1832.]

Land occupied by a servant.

§ 33. The real property of non-resident owners, improved or occupied by a servant or agent, shall be subject to assessment of highway labor, and at the same rate as the real property of resident owners. [*Sec.* 1 *of chap.* 154 *of* 1835.]

Copies of lists.

§ 34. [Sec. 25.] The commissioners shall direct the clerk of the town to make a copy of each list, and shall subscribe such copies; after which, they shall cause the several copies to be delivered to the respective overseers of highways of the several districts in which the highway labor is assessed.

Names omitted, &c.

§ 35. [Sec. 26.] The names of persons left out of any such list, and of new inhabitants, shall from time to time be added to the several lists; and they shall be rated by the overseers in proportion to their real and personal estate, to work on the highways, as others rated by the commissioners on such lists, subject to an appeal to the commissioners.

Appeals by non-residents.

§ 36. [Sec. 27.] Whenever any non-resident owner shall conceive himself aggrieved by the assessments of any commissioners of highways, in carrying into effect the provisions of this Article, it shall be lawful for such owner, or his agent, within thirty days after such assessment, to appeal to any three judges of the Court of Common Pleas of the county in which such land is situated.

Proceedings.

§ 37. [Sec. 28.] It shall be the duty of such judges, within twenty days thereafter, to convene and decide on such appeal, the said owner or agent giving notice to the commissioners of the time of the meeting of the judges; and their decision, or that of any two of them, shall be final and conclusive in the premises. Each judge shall be entitled to receive, for his services on such appeal, two dollars for each day he may be employed thereon, to be paid by the party appealing, if the proceedings of the commissioners and overseers shall be affirmed; but if reversed or modified favorable to the party appealing, to be levied and paid as part of the contingent expenses of such town.

§ 38. [Sec. 29.] It shall be the duty of the commissioners of highways of each town, to credit such persons as live on private roads, and work the same, so much on account of their assessments, as such commissioners may deem necessary to work such private road; or to annex such private roads to some of the highway districts. Private roads.

§ 39. [Sec. 30.] Whenever the commissioners of highways shall assess the occupant, for any land not owned by such occupant, they shall distinguish, in their assessment lists, the amount charged upon such land, from the personal tax, if any, of the occupant thereof; but when any such land shall be assessed in the name of the occupant, the owner thereof shall not be assessed during the same year to work on the highways on account of the same land. Certain assessments to be separate.

§ 40. [Sec. 31.] Whenever any tenant of any land for a less term than twenty-five years, shall be assessed to work on the highways, for such land, pursuant to the last preceding section, and shall actually perform such work, or commute therefor, he shall be entitled to a deduction from the rent due or to become due from him, for such land, equal to the full amount of such assessment, estimating the same at the rate of sixty-two and a half cents per day; unless otherwise provided for, by covenant or agreement, between such tenant and his landlord. Tenant to deduct assessment.

ARTICLE III.

OF THE DUTIES OF OVERSEERS IN REGARD TO THE PERFORMANCE OF LABOR UPON HIGHWAYS, AND OF THE PERFORMANCE OF SUCH LABOR, OR THE COMMUTATION THEREFOR.

§ 41. [Sec. 32.] It shall be the duty of the overseers of highways, to give at least twenty-four hours' notice to all persons assessed to work on the highways, and Notice to work, where to be done.

residing within the limits of their respective districts, of the time and place when and where they are to appear for that purpose, and with what implements; but no person, being a resident of the town, shall be required to work on any highway, other than in the district in which he resides, unless he shall elect to work in some district where he has any land; and in such case he may, with the approbation of the commissioners of highways, apply the work assessed in respect to such land, in the district where the same is situated.

Notice to non-residents.

§ 42. [Sec. 33.] It shall be the duty of the several overseers of highways, to notify the agent of every non-resident landholder, whose lands are assessed (if such agent reside in the town where such assessment is made), of the number of days such non-resident is assessed, and of the time when and the place where the labor is to be performed; which notice shall be given at least five days previous to the time appointed.

Id.

§ 43. [Sec. 34.] If the overseer cannot ascertain that such non-resident has an agent within such town, he shall affix a written notice on the outer door of the building in which the last town meeting in such town was held, containing a list of the names of such non-residents, when known, and a description of the tracts of land comprised in his list, together with the number of days' labor assessed on each tract, and a specification of the time when and the place where such labor is to be performed; which notice shall be posted at least twenty days before the time appointed for performing such labor.

Commutation for work.

§ 44. [Sec. 35.] Every person liable to work on the highways, shall work the whole number of days for which he shall have been assessed; but every such person, other than an overseer, may elect to commute for the same, or for some part thereof, at the rate of sixty-two and a half cents for each day; in which case such commutation money shall be paid to the overseer of highways of the district in which the person commuting shall reside, to be applied and expended by such overseer in the improvement of the roads and bridges in the same district.

§ 45. [Sec. 36.] Every person intending to commute for his assessment, or for any part thereof, shall, within twenty-four hours after he shall be notified to appear and work on the highways, pay the commutation money for the work required of him by such notice; and the commutation shall not be considered as complete, until such money be paid. When to be paid.

§ 46. [Sec. 37.] Every overseer of highways shall have power to require a team; or a cart, wagon or plow, with a pair of horses or oxen, and a man to manage them; from any person having the same within his district, who shall have been assessed three days or more, and who shall not have commuted for his assessment; and the person furnishing the same upon such requisition, shall be entitled to a credit of three days for each day's service therewith. Teams, &c.

§ 47. [Sec. 38.] Every person assessed to work on the highways, and warned to work, may appear in person, or by an able bodied man as a substitute; and the person or substitute so appearing, shall actually work eight hours in each day, under the penalty of twelve and a half cents for every hour such person or substitute shall be in default, to be imposed as a fine on the person assessed. Substitutes, hours to work.

§ 48. [Sec. 39.] If any such person or his substitute shall, after appearing, remain idle, or not work faithfully, or hinder others from working, such offender shall, for every offence, forfeit the sum of one dollar. Penalty for neglect, &c.

§ 49. [Sec. 40.] Every person so assessed and duly notified, who shall not commute, and who shall refuse or neglect to appear as above provided, shall forfeit, for every day's refusal or neglect, the sum of one dollar. If he was required to furnish a team, carriage, man or im- Penalties for not working, &c.

plements, and shall refuse or neglect to comply, he shall be fined as follows:

1. For wholly omitting to comply with such requisition, three dollars for each day;

2. For omitting to furnish a cart, wagon or plow, one dollar for each day;

3. For omitting to furnish a pair of horses or oxen, one dollar for each day;

4. For omitting to furnish a man to manage a team, one dollar for each day.

Complaints how made.

§ 50. [Sec. 41.] It shall be the duty of every overseer of highways, within six days after any person so assessed and notified, shall be guilty of any refusal or neglect for which a penalty or fine is prescribed in this title, unless a satisfactory excuse shall be rendered to him for such refusal or neglect, to make complaint on oath, to one of the justices of the peace of the town.

1 John. 515. 10 do. 470.

Proceedings.

§ 51. [Sec. 42.] The justice to whom such complaint shall be made, shall forthwith issue a summons, directed to any constable of the town, requiring him to summon such delinquent to appear forthwith before such justice, at some place to be specified in the summons, to show cause why he should not be fined according to law for such refusal or neglect; which summons shall be served personally, or by leaving a copy at his personal abode.

3 John. 474. 9 do. 229.

Proceedings.

§ 52. [Sec. 43.] If, upon the return of such summons, no sufficient cause shall be shown to the contrary, the justice shall impose such fine as is provided in this title for the offense complained of, and shall forthwith issue a warrant under his hand and seal, directed to any constable of the town where such delinquent shall reside, commanding him to levy such fine, with the costs of the proceedings, of the goods and chattels of such delinquent.

Id.

§ 53. [Sec. 44.] The constable to whom such warrant shall be directed, shall forthwith collect the moneys therein mentioned. He shall pay the fine when collected, to the justice who issued the warrant, who is hereby required to pay the same to the overseer who entered the complaint, to be by him expended in improving the roads and bridges in the district of which he is overseer.

§ 54. [Sec. 45.] Every penalty collected for a refusal or neglect to appear and work on the highways, shall be set off against the assessment upon which it was founded, estimating every dollar collected as a satisfaction for one day's work. Penalties to be set off.

§ 55. [Sec. 46.] The acceptance by an overseer of any excuse for refusal or neglect, shall not, in any case, exempt the person excused from commuting for, or working, the whole number of days for which he shall have been assessed during the year. Excuses.

§ 56. [Sec. 47.] Every overseer of highways shall, on or before the first day of October, in each year, make out and deliver to the supervisor of his town, a list of all the lands of non-residents, and of persons unknown, which were taxed on his lists, on which the labor assessed by the commissioners of highways has not been paid, and the amount of labor unpaid; and the said overseer, previous to delivering such list, shall make and subscribe an affidavit thereon, before some justice of the peace of such town, that he has given the notice required by the thirty-third [42d] and thirty-fourth [43d] sections of this title, and that the labor for which such land is returned has not been performed. Proceedings to collect non-resident labor unpaid.

§ 57. [Sec. 48.] If any overseer shall refuse or neglect to deliver such list to the supervisor, as provided in the last preceding section, or shall refuse or neglect to make the affidavit as therein directed, he shall for every such offense, forfeit the sum of five dollars, and also the amount of tax or taxes for labor remaining unpaid, at the rate of sixty-two and a half cents for each day; to be recovered by the commissioners of highways of the town, and to be applied by them in making and improving the roads and bridges in such town. Proceedings, &c.

Proceedings, &c. § 58. [Sec. 49.] It shall be the duty of the supervisors of the several towns, to receive the lists of the overseers of highways, when delivered pursuant to the preceding forty-seventh [56th] section, and to lay the same before the board of supervisors of the county.

Id. § 59. [Sec. 50.] It shall be the duty of such board, at their next meeting, to cause the amount of such arrearages of labor (estimating a day's labor at sixty-two and a half cents) to be levied on the lands so returned, and to be collected in the same manner that the contingent charges of the county are levied and collected, and to order the same when collected to be paid over to the commissioners of highways of the town, to be by them applied to the construction and improvement of the roads and bridges in the district for whose benefit the labor was originally assessed.

Annual return of overseers. § 60. [Sec. 51.] Every overseer of highways shall, on the second Tuesday next preceding the time of holding the annual town meeting in his town, within the year for which he is elected or appointed, render to one of the commissioners of highways of the town, an account in writing, verified by his oath [which the commissioners of highways are authorized to administer], and containing:

1. The names of all persons assessed to work on the highways in the district of which he is overseer;

2. The names of all those who have actually worked on the highways, with the number of days they have so worked;

3. The names of all those who have been fined, and the sums in which they have been fined;

4. The names of all those who have commuted, and the manner in which the moneys arising from fines and commutations have been expended by him;

5. A list of all lands which he has returned to the supervisor for non-payment of taxes, and the amount of

tax on each tract of land so returned. [*Amended by chap.* 149 *of* 1833.]

§ 61. [Sec. 52.] Every such overseer shall also then and there pay to the commissioner, all moneys remaining in his hands unexpended, to be applied by the commissioners in making and improving the roads and bridges in the town, in such manner as they shall direct. To pay over moneys.

§ 62. [Sec. 53.] If any overseer shall refuse or neglect to render such account, or if having rendered the same, he shall refuse or neglect to pay any balance which may then be due from him, he shall, for every such offense, forfeit the sum of five dollars, to be recovered with the balance of moneys remaining in his hands, by the commissioners of highways of the town, and to be applied in making and improving the roads and bridges. It shall be the duty of the commissioners of highways to prosecute for such penalty in every instance in which no return is made. Penalty, how collected.

§ 63. Whenever it shall appear from the annual return of any overseer of highways, made in pursuance of the fifty-first section of the sixteenth chapter of title first of the first part of the Revised Statutes, that any person who was assessed to work on the highways (other than non-residents), has neglected to work the whole number of days to him assessed, and has not commuted for or otherwise satisfied such deficiency, then it shall be the duty of the commissioners of highways to reassess such deficiency to the person so delinquent, at the next assessment of work for highway purposes, and to add to it his annual assessment. [*Sec.* 2 *of chap.* 107 *of* 1832.] Reassessment in case of neglect.

§ 64. Such reassessment shall not exonerate any overseer of highways from any penalty which he may have incurred under the sixteenth section of the last aforesaid chapter. [*Sec.* 3, *of same chapter.*] Overseers.

ARTICLE IV.

OF THE LAYING OUT OF PUBLIC AND PRIVATE ROADS, AND OF THE ALTERATION OR DISCONTINUANCE THEREOF.

Who may apply.

§ 65. [Sec. 54.] Every person liable to be assessed for highway labor, may apply to the commissioners of highways of the town in which he shall reside, to alter or discontinue any road, or to lay out any new road. Every such application shall be in writing, addressed to the commissioners, and signed by the person applying.

§ 66. Every person liable to be assessed for highway labor, and owning lands in a town in which he is not a resident, may apply to the commissioners of highways of the town in which the lands are situated, to alter, discontinue, or to lay out any road through the same. [*Chap.* 122 *of* 1836.]

Survey.

§ 67. [Sec. 55.] Whenever the commissioners of highways shall lay out, alter or discontinue any road, either upon application to them or otherwise, they shall cause a survey to be made of such road, and shall incorporate such survey in an order to be signed by them, and to be filed and recorded in the office of the town clerk, who shall note the time of recording the same.

Order to be posted, &c.

§ 68. [Sec. 56.] It shall be the duty of the town clerk, whenever any order of the commissioners for laying out, altering or discontinuing a road shall be received by him, to post a copy of such order on the door of the house where the town meeting is usually held; and the time hereinafter limited for appealing from any such order, shall be computed from the time of recording the same.

Consent of owner, when necessary.

§ 69. [Sec. 57.] No public or private road shall be laid out through any orchard or garden, without the consent of the owner thereof, if such orchard be of the growth of four years or more, or if such garden had been cultivated for four years or more, before the laying out of such road. Nor shall any such road be laid out through

4 Cowen, 190.

any buildings; or any fixtures or erections for the purposes of trade or manufactures; or any yards or enclosures necessary to the use and enjoyment thereof; without the consent of the owner. 4 Paige, 523. 6 do 86. 2 Hill, 443.

§ 70. [Sec. 58.] No highway shall be laid out through enclosed, improved or cultivated land, without the consent of the owner or occupant thereof, unless certified to be necessary by the oath of twelve reputable freeholders of the town, in the manner hereinafter provided. Oath, &c., when necessary. 7 Wend., 264. 3 Hill, 458.

§ 71. [Sec. 59.] Every person who shall apply for the laying out of a highway through any such land, shall cause notices in writing to be posted up at three of the most public places of the town, specifying, as near as may be, the route of the proposed highway, the several tracts of land through which the same is proposed to be laid, and the time and place at which the freeholders will meet to examine the ground. Every such notice shall be posted up at least six days before the time specified therein for the meeting of the freeholders. Notice of application.

§ 72. [Sec. 60.] If twelve reputable freeholders of the town, not interested in the lands through which the road is to be laid, nor of kin to the owner thereof, shall appear at the time and place specified in the notice, they shall then be sworn by [a justice of the peace or] any officer authorized to administer oaths, well and truly to examine and certify, in regard to the necessity and propriety of the highway applied for. Proceedings. 7 Wend., 264.

§ 73. [Sec. 61.] They shall then personally examine the route of such highway, and shall hear any reasons that may be offered for or against laying out the same. If they shall be of opinion that such highway is necessary and proper, they shall make and subscribe a certificate in writing to that effect, which shall be delivered to the commissioners of highways of the town. Proceedings.

§ 74. No compensation shall be allowed any juror for examining and certifying in regard to the necessity and propriety of any highway being laid out, altered or dis- Jurors not to be paid.

continued, nor for appearing to make such examination. [*Sec.* 14 *of chap.* 180 *of* 1845.]

Notice to occupant.

§ 75. [Sec. 62.] Before the commissioners shall determine to lay out the highway, so applied for and certified, they shall cause notice in writing to be given to the occupant of the land through which the road is to run, of the time and place at which they will meet to decide on the application. The notice shall be served by delivering the same to such occupant, or if he be absent, by leaving the same at his dwelling-house; and in either case, at the least three days before the time of meeting.

Description of road.

§ 76. [Sec. 63.] The commissioners shall meet at the time specified in the notice, and shall hear any reasons that may be offered for or against laying out the highway. If they shall determine to lay out such highway, they shall make out and subscribe a certificate of such determination, describing the road so laid out, particularly, by routes and bounds and by its courses and distance, and shall deposit the same with the town clerk.

23 Wend., 326.

§ 77. [Sec. 64.] The damages sustained by reason of the laying out and opening such road may be ascertained by the agreement of the owner and the commissioners of highways, provided such damages do not exceed one hundred twenty-five dollars, and unless such agreement be made, or the owner of the land shall in writing release all claim to damages, the same shall be assessed in the manner prescribed by law, before such road shall be opened, or worked, or used. Every such agreement or release shall be filed in the town clerk's office, and shall forever preclude such owner from all further claim for such damages. [*See* § 22 *of chap.* 455 *of* 1847.]

Damages, how assessed by jury.

§ 78. [Sec. 65.] On the application of the commissioners of highways, or of the owner of the land through which such road is laid out, to any two justices of the peace of the town, they shall issue their warrant to some constable of some other town of the same county, neither interested, nor of kin to any person interested, in the land through which the road is laid out; directing him to summon twelve disinterested freeholders, residing in some other town than that in which such road is laid out, and not of kin to the owner of such land, to assess the damages sustained by the laying out such road; and shall therein specify the time and place at which the jury shall meet. [Modified; see post §§ 82, 83, 84, 85.]

§ 79. [Sec. 66.] Upon such freeholders appearing, the justices who issued the warrant, shall draw, by lot, six of the names of the persons attending, to serve as a jury; and the first six persons drawn, who shall be free from all legal exceptions, shall be the jury to assess the said damages. Proceedings.

§ 80. [Sec. 67.] In all cases of the assessment of such damages, the persons by whom the assessment is to be made, shall view and examine the premises; and before making their determination, the freeholders making the same shall be sworn well and truly to determine and assess such damages. Id.

§ 81. [Sec. 68.] The verdict of the jury assessing such damages, shall be received and certified by the two justices who issued the warrant for summoning them, and shall be delivered by them to the commissioners of highways of the town. Id.

§ 82. Whenever any damages are now allowed to be assessed by law, when any road or highway shall be laid out, altered or discontinued, in whole or in part, such damages shall be assessed by the commissioner of highways, where but one shall be chosen, or the commissioner of the first class, where three shall be chosen, and any two assessors of the town, who shall meet, be sworn and proceed in the same manner as juries are now by law required to be sworn and to proceed; and in case such commissioner, or either of the assessors, shall be incompetent to act in the premises, then his place shall be supplied by a justice of the peace of the town, disinterested in the premises; and whenever an assessment shall have been Damages, how to be assessed.

made under this section, it shall be delivered to said commissioner, who shall file the same with the town clerk within ten days thereafter. [*Sec.* 5 *of chap.* 180 *of* 1845.]

Persons aggrieved may appeal.

§ 83. Any person conceiving himself aggrieved by any such assessment, or the supervisor on the part of the town, may, within twenty days after the filing thereof as aforesaid, signify the same by serving a written notice upon any justice of the peace of the town, stating that such person or supervisor requires a review of such assessment, and that a jury should be called for the purpose; and thereupon such justice shall issue his warrant, and the proceedings shall be had in the same manner as now required by law, in relation to assessments made on application of the commissioners of highways to two justices of the peace; but the cost and expenses of such review shall be paid by the person or persons requiring the same, or by the town, if required by the supervisor. [*Sec.* 6 *of same chapter.*]

ART. 4.

Certificate of amount assessed to be given and audited.

§ 84. In all cases where the assessments of damages for laying out, altering or discontinuing any highway or road, shall be made under either of the two last preceding sections, a certificate of the amount thereof shall be delivered by the supervisor of the town to the board of supervisors of the county, to be audited; and if the supervisor of the town, or any person interested, shall feel aggrieved by such assessment, the same shall, by order of such board, be referred to any three judges of the county for reconsideration, who shall have power to inquire into the principles and fairness of such assessment, and to increase or diminish the damages, as in their judgment shall be just and reasonable. [*Sec.* 7 *of same chapter.*]

May be referred to three judges.

Judges when to meet to review premises and hear parties.

§ 85. Upon receiving notice of such order of reference from the clerk of such board or the party interested therein, the three judges named therein shall meet as soon thereafter as convenient, and review the premises, hear the parties and their proofs, and make their certificate of reassessment, and cause the same to be delivered to the board of supervisors of the county, who shall cause the same to be levied and collected as now required by law, after a final settlement of damages in such cases by such board. [*Sec.* 8 *of same chapter.*]

Damages and expenses, how collected.

§ 86. [Sec. 70.] The amount of damages, as finally settled, together with the charges of the commissioners of highways, justices, surveyors and other persons or officers employed in making the assessment, shall be levied and collected in the town within which the highway shall be situated. The moneys so collected, shall be paid to the commissioners of highways of the same town, who shall pay to the owner the sum assessed to him, and appropriate the residue to satisfy the charges. [*Amended by* §§ 5, 6, 7 *&* 8 *of chap.* 180 *of* 1845, *which also abrogate* § 69, *R. S.*] [*See chap.* 455, *Laws of* 1847.]

12 Wend., 98.

Damages in certain cases, how estimated.

§ 87. [Sec. 71.] Where any person shall be the owner of any land over which any highway shall run, and such highway shall be discontinued, in whole or in part, by reason of some other road to be established and laid out under this title, through the lands of the same person, the persons who shall assess the damages shall take into calculation the value of the road so discontinued, and the benefit resulting to such person by reason of such discontinuance, and shall deduct the same from the damages assessed for the opening and laying out of such new road; and thereupon the owner of the land may enclose so much of the highway so discontinued, as shall belong to him.

Disagreement respecting certain roads.

§ 88. [Sec. 72.] When the commissioners of highways of any town shall disagree with the commissioners of any other town in the same county, relating to the laying out of a new road, or the alteration of an old road, extending into both towns; or when the commissioners of a town in one county shall disagree with the commissioners of a town in another county, relative to laying out a new road, or altering an old road, which shall extend into both counties; the commissioners of both towns shall meet together at the request of either disagreeing commissioners, and make their determination upon such subject of disagreement.

Road upon line of two towns.

§ 89. [Sec. 73.] Whenever it shall become necessary to have a highway upon the line between two towns, such highway shall be laid out by two or more of the commissioners of highways of each of said towns, either upon such line, or as near thereto as the convenience of the ground will admit; and they may so vary the same either to the one or to the other side of such line, as they may think proper.

How divided into districts.

§ 90. [Sec. 74.] It shall be the duty of the same commissioners, when they lay out such highway, to divide it into two or more road districts, in such manner that the labor and expense of opening, working, and keeping in repair such highway, through each of the said districts, may be equal as near as may be, and to allot an equal number of the said districts to each of the said towns.

Effect of allotment.

§ 91. [Sec. 75.] Each district shall be considered as wholly belonging to the town to which it shall be allotted, for the purpose of opening and improving the road, and for keeping it in repair; and the commissioners shall cause such highway, and the partition and allotment thereof, to be recorded in the office of the town clerk in each of their respective towns.

Former roads.

§ 92. [Sec. 76.] All highways heretofore laid out upon the line between any two towns, shall be divided, allotted, recorded, and kept in repair, in the manner above directed.

Private roads, how laid out.

24 Wend., 367; 4 Hill, 150; 6 do. 47.

§ 93. [Sec. 77.] Whenever application shall be made to the commissioners of highways of any town, for a private road, they shall summon twelve disinterested freeholders of the town where the land through which such road is proposed to be laid out, is situated, to meet on a day certain; of which day notice shall be given to the owner or occupant of such land. Such freeholders, when met, shall be sworn as above provided, and shall then proceed to view the lands through which such road is applied for.

§ 94. [Sec. 78.] If they shall determine that such road is necessary, they shall make and subscribe a certificate in manner aforesaid, and the commissioners shall thereupon lay out the road, and cause a record thereof to be made in the town clerk's office. The damages of the owner of the land through which such road shall be laid out, shall be ascertained or assessed in like manner as if the same was a public highway, and such damages shall be paid by the person applying for the road. [*See* § 10, *chap.* 455 *of* 1847.]

Proceedings.

10 Wend., 585.

§ 95. [Sec. 79.] Every such private road, when so laid out, shall be for the use of such applicant, his heirs and assigns; but not to be converted to any other use or purpose than that of a road. Nor shall the occupant or owner of the land through which such road shall be laid out, be permitted to use the same as a road, unless he shall have signified his intention of so making use of the same to the jury or commissioners who ascertained the damages sustained by laying out such road, and before such damages were so ascertained.

For what purpose road to be used.

14 John., 383.

5 Wend., 580.

§ 96. [Sec. 80.] All public roads to be laid out by the commissioners of highways of any town, shall not be less than three rods wide, and all private roads shall not be more than three rods wide.

Width of roads.

§ 97. [Sec. 81.] Whenever application shall be made for the discontinuance of an old road, on the ground that it has become useless and unnecessary, the commissioners of highways to whom such application shall be made, shall summon twelve disinterested freeholders of the town, to meet on a day certain, to consider such application. Such freeholders, when met, shall be sworn well and truly to examine and certify in regard to the propriety of such discontinuance.

Old roads, how discontinued.

§ 98. [Sec. 82.] They shall then proceed to view such road, and if they shall be of opinion that the

Id.

5

same is useless and unnecessary, they shall make and subscribe a certificate in writing to that effect, which shall be delivered to the commissioners of highways, who shall thereupon proceed to decide upon such application.

Papers, where filed.

§ 99. [Sec. 83.] All applications, certificates and other papers relating to the laying out, altering or discontinuing of any road, shall be filed by the commissioners of highways, as soon as they shall have decided thereon, in the office of the town clerk of the town.

Appeals.

§ 100. [Sec. 84.] Every person who shall conceive himself aggrieved by any determination of the commissioners of highways, either in laying out, altering or discontinuing any road, may, at any time within sixty days thereafter, appeal to any three of the judges of the court of common pleas of the county in which such road is situated; but an appeal by one person, and a decision thereon, shall not conclude nor affect the rights of any other person who shall appeal within the limited period. [*Amended as to refusing to lay out, &c., and other provisions made by chap.* 180 *of* 1845, *see post* §§ 113 *to* 117]

1 Paige, 521; 15 John., 537; 24 Wend., 491; 25 do. 453.

Power and duty of judges.

§ 101. [Sec. 85.] The judges to whom the first appeal from any such determination shall be made, shall have exclusive jurisdiction of all appeals from the same determination, to the end that their decision, when made, may embrace the whole subject; and for this purpose they shall suspend all proceedings upon the appeal first made, and upon all other appeals received by them from such determination, until the time limited for such appeals shall have expired.

7 Wend., 265.

Right of appeal.

§ 102. [Sec. 86.] Every such appeal shall be in writing, addressed to the judges, and signed by the party appealing. It shall briefly state the ground upon which it is made, and whether it is brought to reverse entirely the determination of the commissioners, or only to reverse a part thereof; and in the latter case, it shall specify what part. [*Amended by* § 8, *chap.* 455 *of* 1847.]

§ 103. [Sec. 87.] It shall be the duty of the judges to whom the appeal is made, to proceed thereon as soon as may be convenient. Where the determination appealed from was against an application for laying out, altering or discontinuing a road, the judges shall give notice to the commissioners by whom such determination was made. Where the appeal is from a determination in favor of an application for laying out, altering or discontinuing a road, the notice shall be given to the commissioners, and to one or more of the applicants for such road. In all cases the notice shall specify the time and place at which the judges will convene to hear the appeal. [*See* § 8, *chap.* 455 *of* 1847.] Proceedings. 20 Wend., 186.

§ 104. [Sec. 88.] Every such notice shall be served at least eight days before the time mentioned therein, by delivering the same to one of the commissioners whose determination is appealed from, or by leaving the same at his dwelling-house. If the notice be also directed to an applicant, it shall be served in the same manner. Notice of appeal.

§ 105. [Sec. 89.] It shall be the duty of the judges to convene at the time and place mentioned in the notice, and to hear the proofs and allegations of the parties. They shall have power to issue process to compel the attendance of witnesses, and may adjourn from time to time, as may be necessary. Their decision, or that of any two of them, shall be conclusive in the premises; and every such decision shall be reduced to writing, be signed by the judges making it, and be filed by them in the office of the town clerk of the town, who shall record the same. [*See* § 8, *chap.* 455 *of* 1847.] Proceedings. 2 Caines, 179; 15 John., 537; 13 Wend., 432; 25 do. 453.

§ 106. [Sec. 90.] Every such judge shall be entitled to receive two dollars for every day employed in the hearing and decision of such appeal, to be paid by the party appealing where the determination of the commissioners shall be affirmed; but where it is reversed, to be a charge against the county. Fees.

Vacancies. § 107. [Sec. 92.] In case the office of any one of the judges to whom such appeal shall be made, shall become vacant before the determination of such appeal, it shall be the duty of the remaining judges named therein, to associate with themselves another of the judges of the same court, who shall act with them in all subsequent proceedings in the same manner as if he had been originally named in such appeal.

How altered. § 108. [Sec. 93.] No road which has been fixed by the decision of the judges on an appeal to them, shall be discontinued or altered so long as such judges, or either of them, shall continue in commission, except by the order of the same judges, or such of them as continue in commission, joined with such other judge or judges as shall be necessary to make three, such additional judge or judges to be selected by the person applying for the discontinuance or alteration.

§ 109. [Sec. 94.] If no one of the said judges shall continue in commission, such application shall be made to any three of the judges of the same court, not having any interest in the road so desired to be discontinued or altered.

§ 110. [Sec. 95.] No application made under either of the two last preceding sections shall be acted upon by the judges, unless the same be accompanied by a certificate signed by the commissioners of highways of the town in which the road is situated stating their approbation of such application; and before the judges decide thereon, they shall proceed to view the road so desired to be discontinued or altered. They shall be entitled to the same compensation as above provided, to be paid by the applicant.

Fences to be removed. § 111. [Sec. 96.] Whenever the commissioners of highways shall have laid out any public highway, through any inclosed, cultivated or improved lands, in conformity to the provisions of this title, and their determination shall not have been appealed from, they shall give the owner or occupant of the land through which such road shall have been laid, sixty days' notice, in writing, to remove his fences. If such owner shall not remove his fences within the sixty days, the commissioners shall cause such fences to be removed, and shall direct the road to be opened and worked.

§ 112. [Sec. 97.] If the determination of the commissioners shall have been appealed from, then the sixty days' notice shall be given, after the decision of the referees upon such appeal shall have been filed in the office of the town clerk of the town. [*Amended by* § 8 *of chapter* 455 *of* 1847.] Fences to be removed.

§ 113. Whenever any appeal shall hereafter be made from any decision of any highway commissioner or commissioners for refusing to lay out, alter or discontinue any road or highway, such appeal shall in the first instance be made to the first judge of the county courts of the county wherein such commissioner shall reside; or in case of a vacancy in the office of such judge, or in case of his interest or disability, then to any other disinterested county judge of such county; and such appeal shall be brought and conducted in all respects as appeals in like cases are now required to be by law; provided, that any judge being a resident of the town where such road or highway shall be located, shall be deemed interested in the matter so as to prevent his acting on any such appeal. [*See Laws of* 1847.] Appeal to be made to first judge.

§ 114. Any party or person conceiving himself aggrieved by any decision upon any appeal under the last preceding section, may, at any time within forty days thereafter, file a notice in the office of the clerk of the town where the highway commissioner or commissioners shall reside, signifying that he intends to appeal from the decision of such judges to two other judges of the same county, who shall be named in such notice to be associated with the person who made the decision on the first appeal. The party who shall thus file a notice, shall, within ten days thereafter, serve a copy of such notice upon Persons aggrieved may appeal to two other judges.

each of said judges, and also such other notices as are required to be given on appeals, as now conducted, from commissioners of highways, to three county judges, of the hearing of appeals. [*See Laws of* 1847.]

Judges to hear appeals, &c.; their decision final.

§ 115. The judges associated together under the tenth section of this act, shall entertain and hear all appeals in relation to the same matter, meet and determine the same and file their decision in the office of the clerk of the town as soon as convenient after the expiration of the said forty days, and their decision shall be final. [*See Laws of* 1847.]

Final determination to be carried into effect by commissioners.

§ 116. Whenever there shall have been any final determination upon any appeal or appeals provided for as aforesaid, making it necessary that any road or highway shall be laid out, altered, opened or discontinued, it shall be the duty of the commissioner or commissioners of highways of the town where the same is to be done, to carry out such determination the same as if the decision of such commissioner or commissioners had been in favor thereof, and there had been no appeal. [*Sec.* 13, *chapter* 180 *of* 1845.]

Costs to be paid by appellant.

§ 117. The party or persons appealing in pursuance of section ten of this act, shall pay all costs thereof, at the same rates as are now chargeable in cases of appeals from the decisions of highway commissioners, whether the decision upon such appeal shall be reversed, affirmed or modified. [*See Laws of* 1847.]

Certain acts of commissioners confirmed.

§ 118. [Sec. 98.] The acts and doings of the commissioners of highways of the several towns in this state, or of any two of them, in laying out, altering or discontinuing any road or highway, since the thirty-first day of December, one thousand eight hundred and five, and prior to the fourteenth day of April, one thousand

7 Wend., 145.

eight hundred and twenty-six, are confirmed from the last mentioned day, provided such commissioners, or any two of them, shall have caused a survey of such roads or highways to be filed and recorded in the office of the town clerk of the town. But such confirmation shall not affect any decision of the judges of the court of common pleas, made prior to the fourteenth day of April, one thousand eight hundred and twenty-six, confirming or reversing the determination of the said commissioners; nor any appeal from such determination, made within six months after that day; nor any suits or proceedings which on that day were pending, at law or in equity.

§ 119. [Sec. 99.] Every public highway already laid out, that shall not have been opened and worked within six years from the time of its being so laid out, and every such highway hereafter to be laid out, that shall not be opened and worked within the like period, shall cease to be a road for any purpose whatever. When roads cease. 2 Cowen, 426.

§ 120. [Sec. 100.] All public highways now in use, heretofore laid out and allowed by any law of this state, of which a record shall have been made in the office of the clerk of the county or town; and all roads not recorded, which have been or shall have been used, as public highways, for twenty years or more; shall be deemed public highways, but may be altered in conformity to the provisions of this Title. What roads highways. 7 John., 106; 2 do. 424; 17 do. 277.

§ 121. [Sec. 101.] It shall be the duty of the commissioners of highways to order the overseers of highways to open all roads to the width of two rods at least, which they shall judge to have been used as public highways for twenty years. Width.

§ 122. Whenever any turnpike corporation shall become dissolved, or the road discontinued, its road shall become a public highway, and be subject to all the legal provisions regulating highways. [*Sec.* 1 *of chap.* 262 *of* 1838.] When turnpike roads to become public highways.

ARTICLE V.

REGULATIONS AND PENALTIES CONCERNING THE OBSTRUCTION OF HIGHWAYS, AND ENCROACHMENTS THEREON.

Penalty for obstructing 9 John., 349, 365; 23 Wend., 451.

§ 123. [Sec. 102.] Whoever shall obstruct any highway, or shall fill up or place any obstruction in any ditch constructed for draining the water from any highway, shall forfeit for every such offense the sum of five dollars.

Fences, when and how to be removed.

2 Cowen, 424; 9 John., 359; 6 Wend. 634; 7 do. 309; 14 do. 254; 2 Hill, 472.

§ 124. [Sec. 103.] In every case where a highway shall have been laid out, and the same has been or shall be encroached upon by fences, erected by any occupant of the land through or by which such highway runs, the commissioners of highways of the town shall, if in their opinion it be deemed necessary, order such fences to be removed, so that such highway may be of the breadth originally intended. The commissioners making the order, shall cause the same to be reduced to writing and signed. They shall also give notice in writing, to the occupant of the land, to remove such fences within sixty days. Every such order and notice shall specify the breadth of the road originally intended, the extent of the encroachment, and the place or places in which the same shall be.

Penalty.

§ 125. [Sec. 104.] If such removal shall not be made within sixty days after the service of such notice, the occupant to whom the notice shall be given, shall forfeit the sum of fifty cents for every day, after the expiration of that time, for which such fences shall continue unremoved.

Expense of removal.

And the commissioners of highways may remove or cause to be removed such encroachment, and the occupant of the premises shall pay to the commissioners of highways all reasonable charges therefor, to be collected in the manner provided in the forty-fifth section of said title. [*Amended by* § 1 *of chapter* 300 *of* 1845.]

§ 126. [Sec. 105.] If the occupant to whom notice is given, shall [within five days] deny such encroachment, the commissioners, or some of them, shall apply to any justice of the peace of the county, for a precept directed to any constable of the town, to summon twelve freeholders thereof, to meet at a certain day and place, to be specified in such precept, and not less than four days after the issuing thereof, to inquire into the premises. The constable to whom such precept shall be directed, shall give at least three days' notice to the commissioners of highways of the town, and to the occupant of the land, of the time and place at which such freeholders are to meet. [*Amended by* § 2 *of same chapter.*]

Proceedings if encroachment be denied.

13 John., 460; 3 Wend., 468.

§ 127. [Sec. 106.] On the day specified in the precept, the jury so summoned shall be sworn by such justice well and truly to inquire whether any such encroachment has been made, and by whom. Such witnesses as may be produced by either party, shall also be sworn by such justice; and the jury shall hear the proofs and allegations which may be produced and submitted.

Id.

§ 128. [Sec. 107.] If the jury find that any encroachment has been made, they shall make and subscribe a certificate in writing, stating the particulars of such encroachment, and by whom made, which shall be filed in the office of the town clerk. The occupant of the land, whether such encroachment shall have been made by him, or by any former occupant, shall remove his fences within sixty days after the filing of such certificate, under the penalty provided in the one hundred and fourth [125th] section of this title. He shall also pay the costs of such inquiry; and if the same shall not be paid within ten days, the justice shall issue a warrant for the collection thereof, in the manner provided in the forty-third [52d] section of this title.

Verdict, how enforced.

22 Wend., 134.

7 Wend., 300.

§ 129. [Sec. 108.] If the jury find that no encroachment has been made, they shall so certify, and shall also ascertain and certify the damages which the then

Id.

occupant shall have sustained by such proceeding; which, together with the costs thereof, shall be paid by the commissioners, and shall be a charge in their favor against the town by which they shall have been elected.

When fences to be removed.

§ 130. [Sec. 109.] No person shall be required to remove any fence under the preceding provisions of this article, except between the first day of April and the first day of November in any year.

Penalty for injuring sidewalks.

§ 131. It shall be lawful for any person owning or occupying lands adjoining a highway or road, to construct a sidewalk within such highway or road, and along the line of such land, and also to erect a railing thereto; and when a sidewalk shall be so constructed, every person who shall ride or drive a horse or team upon it, shall forfeit the sum of one dollar, to the use of such owner or occupant, to be sued for in any court having cognizance thereof. [*Sec.* 1 *of chap.* 281 *of* 1836.]

Saving clause.

§ 132. This act shall not be so construed as to diminish, or in any wise interfere with the authority of the overseers of highways, or any other authority legally exercised over highways or roads. [*Sec.* 2 *of same chapter.*]

Fallen trees to be removed.

§ 133. [Sec. 110.] If any tree shall fall, or be fallen by any person, from any inclosed land into any highway, any person may give notice to the occupant of the land from which such tree shall have fallen, to remove the same within two days. If such tree shall not be removed within that time, but shall continue in such highway, the occupant of the land shall forfeit the sum of fifty cents for every day thereafter until such tree shall be removed.

Penalty for cutting trees.

§ 134. [Sec. 111.] In case any person shall cut down any tree on land not occupied by him, so that it shall fall into any highway, river or stream, unless by the order and consent of the occupant, the person so offending shall forfeit to such occupant the sum of one dollar for every tree so fallen, and the like sum for every day the same shall remain in such highway, river or stream.

For not removing from streams.

§ 135. [Sec. 112.] Whoever shall cut, or cause to be cut down, any tree, so that the same shall fall into any river or stream which now is or hereafter shall be declared a public highway, and shall not remove the same out of such river or stream within twenty-four hours thereafter, shall forfeit five dollars for every tree so cut down and left remaining.

Swinging gates.

§ 136. [Sec. 113.] No swinging or other gates shall be allowed on any public highway, laid out by virtue of this title, or which has heretofore been laid out, other than such public highways as run through lands liable to be overflowed by the waters of the adjacent rivers or streams, in such manner as to remove the fences thereon.

How erected and preserved.

§ 137. [Sec. 114.] Such gates shall be erected and kept in good repair, by the overseers of highways of the town, at the proper costs and charges of the occupant of the land, for whose benefit the same shall be erected.

Expense.

§ 138. [Sec. 115.] If more than one gate shall be erected, and the intermediate land between the gates, at the extremities of such lands, shall be in the occupation of more than one person benefited by such gates, the whole charge of erecting and keeping the same in repair shall be borne by all the occupants benefited thereby, in proportion to the extent of land each occupies adjoining the highway, between the gates at the extremities aforesaid.

Proceedings to collect.

§ 139. [Sec. 116.] The overseer of every road district in which such gates shall be, shall, on or before the first day of November in every year, make out and file with the town clerk a statement of the charges incurred in the erection or repairing of such gates, with the name of the person bound to defray the same; which account shall be verified by the oath of such overseer. If more than one person is liable to defray such charges, the statement shall also contain an apportionment thereof between such persons, stating the amount to be paid by each.

Proceedings to collect.

§ 140. [Sec. 117.] The overseers shall, within ten days after filing the statement, demand of every person bound to pay such charges, or to contribute thereto, the sum due from him according to such statement; and if any person shall refuse or neglect to pay such moneys within six days after demand, it shall be the duty of the overseer to make complaint to a justice of the peace of the town, and the like proceedings shall be had for the recovery of such moneys, as in recovery of fines for refusing or neglecting to work on the highways.

Gates to be closed, &c., penalty.

§ 141. [Sec. 118.] The commissioners of highways shall file an account of such gates in the town clerk's office; and if any person shall open any such gate, and shall not, immediately after having passed the same, close it, or shall willfully or unnecessarily ride over any of the grounds adjoining the road on which such gates shall be permitted, he shall forfeit to the party injured treble damages.

ARTICLE VI.

OF THE ERECTION, REPAIRING AND PRESERVATION OF BRIDGES.

When bridges at expense of county.

§ 142. [Sec. 119.] Whenever it shall appear to the board of supervisors of any county, that any one of the towns in such county would be unreasonably burdened by erecting or repairing any necessary bridge or bridges in such town, such board of supervisors shall cause such sum of money to be raised and levied upon the county, as will be sufficient to defray the expenses of erecting or repairing such bridge or bridges, or such part of such expenses as they may deem proper; and such moneys, when collected, shall be paid to the commissioners of highways of the town in which the same are to be expended.

§ 143. [Sec. 120.] No board of supervisors shall, under the last preceding section, cause any sum exceeding one thousand dollars, to be levied and raised on any county in any one year. Limit.

§ 144. [Sec. 121.] In case the commissioners of highways of any town shall be dissatisfied with the determination of the board of supervisors of their county, touching an allowance for any such bridges, such determination shall, on the application of the commissioners, be reviewed by the court of sessions of the same county, whose order in the premises shall be observed by every such board of supervisors. [*Amended by* § 15 *of chap.* 455 *of* 1847.] Provision respecting allowance made by the supervisors.

§ 145. Whenever any adjoining town shall be liable to make and maintain any bridges over any stream dividing such towns, such bridges shall be built and repaired at the equal expense of said towns, without reference to the town lines. [*Sec.* 1 *of chap.* 225 *of* 1841.] Bridges, how to be built and repaired between towns.

§ 146. For the purpose of building and repairing such bridges, it shall be lawful for the commissioners of highways in said adjoining towns to enter into joint contracts, and such contracts may be enforced in law or equity against such commissioners jointly, the same as if entered into by individuals; and said commissioners may be proceeded against jointly for any neglect of duty in reference to such bridges. [*Sec.* 2 *of same chapter.*] Joint contract to be entered into by commissioners.

§ 147. If the commissioners of highways of either of such towns, after reasonable notice in writing from the commissioners of highways of any other of such towns, shall neglect or refuse to build or repair any such bridge, it shall be lawful for the commissioners so giving such notice, to make or repair such bridges, and then to maintain a suit at law in their official capacity against said commissioners so neglecting or refusing to join in such making or repairing; and in such suit, the plaintiff shall be entitled to recover one-half of the expense of such building or repairing, with costs of suit and interest, without proving any contract. [*Sec.* 3 *of same chapter.*] Proceedings in case of neglect or refusal of either town.

Judgment recovered to be a charge on town; how collected.

§ 148. Any judgment recovered against the commissioners of highways in their official capacity under the provisions of this act, shall be a charge on said town, and collected in the same manner as other town charges, except in cases where the court before which the judgment shall be recovered shall certify that the neglect or refusal of said commissioners was willful and malicious, in which case said commissioners shall be personally liable for such judgment, and the same may be enforced against them in the same manner as against individuals. [*Sec.* 4 *of same chapter.*]

Notice of fine, &c.

§ 149. [Sec. 122.] The commissioners of highways of each town may put up and maintain in conspicuous places, at each end of any bridge in such town, maintained at the public charge, and the length of whose cord is not less than twenty-five feet, a notice with the following words in large characters: "One dollar fine for riding or driving on this bridge faster than a walk."

Penalty.

§ 150. [Sec. 123.] Whoever shall ride or drive faster than on a walk, over any bridge, upon which such notices shall have been placed, and shall then be, shall forfeit for every offense the sum of one dollar.

Injuries to bridges.

§ 151. [Sec. 124.] Whoever shall injure any bridge maintained at the public charge, shall, for every offense, forfeit treble damages.

Notice at toll bridges.

§ 152. It shall be lawful for any corporation or individual owning a toll bridge, to put up at each end thereof, in a conspicuous place, a notice in the following words, in large characters: "One dollar fine for riding or driving faster than a walk on this bridge:" and during the continuance of such notice, any person who shall ride or drive faster than a walk on such bridge, shall forfeit the sum of one dollar, to be sued for in the name of the corporation or person or persons owning such bridge, and to be recovered with costs of suit. [*Sec.* 2 *of chap.* 262 *of* 1838.]

§ 153. Whenever a corporation owning a toll bridge shall become dissolved, such bridge shall be left without waste or damage, and shall be a public highway. [*Sec.* 3 *of same chapter.*] When to become part of highway.

ARTICLE VII.

MISCELLANEOUS PROVISIONS OF A GENERAL NATURE.

§ 154. [Sec. 125.] Any two commissioners of highways of any town, may make any order in execution of the powers conferred in this title; provided it shall appear in the order filed by them, that all the commissioners of highways of the town met and deliberated on the subject embraced in such order, or were duly notified to attend a meeting of the commissioners, for the purpose of deliberating thereon. Two commissioners may act. 22 Wend., 134.

§ 155. Whenever any association or individual shall construct a railroad upon land purchased for that purpose, on a route which shall cross any road or other public highway, it shall be lawful for the commissioners of highways having the supervision thereof, to give a written consent that such railroad may be constructed across or on such road or other public highway; and thereafter such association or individual shall be authorized to construct and use such railroad across or on such roads or other highways, as the commissioners aforesaid shall have permitted; but any public highway thus intersected or crossed by a railroad, shall be so restored to its former state, as not to have impaired its usefulness. [*Chap.* 300 *of* 1835.] Consent to railroads crossing highway.

§ 156. [Sec. 126.] All trees standing or lying on any land over which any highway shall be laid out, shall be for the proper use of the owner or occupant of such land, except such of them as may be requisite to make or repair the highways or bridges on the same land. Trees; to whom they belong.

§ 157. [Sec. 127.] Any person owning land adjoining any highway not less than three rods wide, may plant or Trees may be planted.

set out trees on the side of such highway contiguous to his land; which trees shall be set in regular rows, at a distance of at least six feet from each other. Whoever shall cut down, destroy or injure any tree that has been or shall be so planted or set out, shall be liable in damages to the owner of such adjoining land.

Penalty for injuring.

Penalty for injuring mile-boards.

§ 158. [Sec. 128.] Whoever shall destroy, remove, injure or deface any mile-board or mile-stone, erected on any highway, shall forfeit for every offense the sum of ten dollars; he shall also be deemed guilty of a misdemeanor, and on conviction shall be fined not exceeding fifty dollars, or imprisoned not exceeding three months, at the discretion of the court.

Penalty for injuring guide-posts.

§ 159. [Sec. 129.] Whoever shall injure or deface any description affixed to a guide-post erected on any highway, or destroy or injure any such guide-post, shall be liable to all the penalties provided in the last preceding section.

Injuries to road.

23 Wend., 451.

§ 160. [Sec. 130.] Whoever shall injure any highway, by obstructing or diverting any creek, water-course or sluice, or by drawing logs of timber on the surface of any road or bridge, or by any other act, shall, for every such offense, forfeit treble damages.

Penalties, how recovered.

5 Hill, 251.

§ 161. [Sec. 131.] All penalties or forfeitures given in this title, and not otherwise specially provided for, shall be recovered by the commissioners of highways of the town in which the offense shall be committed; and when recovered, shall be applied by them in improving the roads and bridges in such town.

Extent of this title.

§ 162. [Sec. 132.] The provisions of this title shall be construed to extend to all parts of this state, except where special provisions inconsistent therewith, have been or shall be made by law, in relation to particular counties, cities, villages or towns.

§ 163. The first title of the sixteenth chapter of part first of the Revised Statutes, shall, immediately on the passing of this act, be in force in the county of Richmond. [*Sec.* 2 *of chap.* 97 *of* 1833.]

OTHER PROVISIONS OF LAW

RELATING TO THE ELECTION, QUALIFICATION, AND DUTIES OF COMMISSIONERS OF HIGHWAYS.

Chap. xi, *Tit.* 2, *Art.* 1, *Revised Statutes.*

The electors to determine whether one or three highway commissioners shall be chosen.

§ 6. The electors of each town shall have power at their annual town meeting, to determine by resolution whether there shall be chosen one or three highway commissioners, and the number so determined upon shall be balloted for and chosen; and if only one shall be determined upon and chosen, he shall possess all the powers and discharge all the duties of commissioners of highways as prescribed by law, and shall hold his office for one year. And whenever three commissioners shall be chosen in any town, they shall be divided by lot by the canvassers, upon the result of the canvass, into three classes, to be numbered one, two, and three; the term of office of the first class shall be one year, of the second, two, and of the third, three; and one commissioner only shall thereafter annually be elected in such town, who shall hold his office for three years, and until a successor shall be duly elected or appointed; but in case any commissioner shall be elected to fill a vacancy, he shall hold the office only for the unexpired term which shall have become vacant; and if two vacancies shall be required to be filled, the canvassers shall, after the canvass, determine by lot as aforesaid, the terms they shall respectively hold. And when any vacancy shall happen by death, removal, resignation, neglect to qualify, or refusal to serve, it shall be supplied until the next succeeding annual town meeting by an apportionment in writing, under the hands of any three justices of the peace, or

Their powers and term of office.

two justices and the supervisor of the town; and every commissioner of highways shall be authorized to administer oaths to any witnesses or juries, in proceedings which may be had by or before them; and whenever any town shall have determined upon having three commissioners, and shall desire to return two, or have but one, such town shall have the power so to do by a resolution taken at an annual town meeting, and when such resolution shall have been adopted, no other commissioner shall be elected or appointed, until the term or terms of those in office at the time of adopting such resolution shall expire or become vacant; and they shall have power to act until their terms shall severally become vacant or expire, as fully as if the three continued in office. [*This section appears here as amended by* § 1 *of chap.* 455 *of* 1847.]

Application for additional tax.

§ 10. Whenever the commissioners of highways of any town in this state, shall be of opinion that the sum of two hundred and fifty dollars, as now allowed by law, will be insufficient to pay the expenses actually necessary for the improvement of roads and bridges, it shall be lawful for such commissioners to apply, in open town meeting, for a vote authorizing such additional sum to be raised as they may deem necessary for the purpose aforesaid, not exceeding two hundred and fifty dollars, in addition to the sum now allowed by law. [*Sec.* 1 *of chap.* 274 *of* 1832.]

Notice, how to be given.

§ 11. Before making such application, it shall be the duty of the commissioners to give notice of their intended application, by posting the same in a conspicuous manner in at least five of the most public places in such town, at least four weeks next preceding the annual town meeting. Such notice shall specify the amount to be applied for, and the purposes for which the same is intended to be appropriated, with the probable amount necessary to be expended at each place, if there shall be more than one. [*Sec.* 2 *of same chapter.*]

Accounts to be exhibited.

§ 12. Whenever any application for a grant of money for the purposes mentioned in the first section of this act, shall be made to any town meeting, it shall be the duty of the commissioners making the same to exhibit a statement of their accounts, and an estimate of the expenses necessary for the improvement of roads and bridges in such town the ensuing year. [*Sec.* 3 *of same chapter.*]

§ 13. If the town meeting shall, by their votes, determine that a sum over and above the amount now allowed by law will be necessary for the improvement of roads and bridges, or to pay any balance that may be due, the clerk shall enter such resolution as shall be agreed to, in the minutes of the meeting, and deliver a copy thereof to the supervisor of the town, who shall lay the same before the board of supervisors at their next annual meeting; and it shall be their duty to cause the amount specified in such resolution to be levied and collected, in the same manner as other town charges of such town. [*Sec.* 4 *of same chapter.*] Order to levy tax.

Title 3. *Article* 2.

§ 42. Every commissioner of highways hereafter to be elected or appointed, shall, before entering upon his duties, and within ten days after notice of his election or appointment, execute to the supervisor of his town, a bond with two sureties, to be approved by the supervisor by an indorsement thereon, and filed with him, in the penal sum of one thousand dollars, conditioned that he will faithfully discharge his duties as such commissioner, and within ten days after the expiration of his term of office, pay over to his successor what money may be remaining in his hands as such commissioner, and render to such successor a true account of all moneys received and paid out by him as such commissioner. [*Sec.* 3 *of chap.* 180 *of* 1845.] Commissioners of highways to give security.

§ 44. [Sec. 25.] If any person chosen or appointed to the office of supervisor, town clerk, assessor, commissioner of highways, or overseer of the poor, shall refuse to serve, he shall forfeit to the town the sum of fifty dollars. Penalties therefor.

TITLE II.

OF THE REGULATION OF FERRIES.

Licenses, by whom granted. 11 Wend., 590. § 1. The court of common pleas in each of the counties of this state, shall grant licenses for keeping ferries in their respective counties, to as many suitable persons as they may think proper; which licenses shall continue in force for a term to be fixed by the court, not exceeding three years.

To whom. § 2. No such license shall be granted to any person other than the owner of the land through which the highway adjoining to the ferry shall run, unless such owner shall neglect to apply for such license, after notice as hereinafter provided.

Id. § 3. Whenever application for a ferry shall be made by any person other than such owner, the court shall not grant a license to such applicant, unless proof shall be made that the applicant caused notice in writing to be given to such owner, at least eight days before the sitting of the court, of his intention to make such application.

Recognizance. § 4. Every person applying for such license, shall, before the same be granted, enter into a recognizance to the people of this state, in open court, in the sum of one hundred dollars, faithfully to keep and attend such ferry, with such and so many sufficient and safe boats, and so many men to work the same, as shall be deemed necessary, together with sufficient implements for said ferry, during the several hours in each day, and at such several rates as the court granting the license, shall, from time to time order and direct; which recognizance shall be forthwith filed with the clerk of the county.

§ 5. Every license so granted, shall be entered in the book of minutes of the court, by the clerk; and a copy thereof, attested by him, shall be delivered to the person licensed. To be entered.

§ 6. Whenever the waters over which any ferry may be used, shall divide two counties, a license obtained in either of the said counties, shall be sufficient to authorize the person obtaining the same, to transport persons, goods, wares and merchandise, to and from either side of said waters. Effect of certain licenses.

§ 7. Every person who shall violate the condition of such recognizance, shall be considered guilty of a misdemeanor; and on conviction, shall be subject to such fine as the court may adjudge not exceeding twenty-five dollars for each offense; and on proof of such conviction, the court of common pleas shall direct the recognizance entered into by such person, to be estreated for the use of the people of this state. Penalty for misconduct.

§ 8. If any person (except within the counties of Essex and Clinton, the counties of Orange, Rockland and Westchester, and the counties in the first senate district), shall use any ferry for transporting across any river, stream, or lake, any person, or goods, or chattels or effects, for profit or hire, unless authorized in the manner directed in this title, such person shall be considered guilty of a misdemeanor; and on conviction, shall be subject to such fine, for the use of the county, as the court may adjudge, not exceeding twenty-five dollars for each offense. Penalty for ferrying without license. 5 John., 175; 11 Wend., 590.

§ 9. Where any such offense shall be committed on waters dividing two counties, the person so offending may be proceeded against in each of said counties; but the fine to be imposed, shall not exceed twelve dollars and fifty cents in each case. Proceedings.

Limitation of this title.

§ 10. Nothing in this title contained, shall affect or alter the ferries granted by charter to the corporations of Albany and Hudson, or alter or impair any grants made by this state, or any legal right or privilege whatever, belonging to any individual or corporation, by virtue of any law of this state, or otherwise.

NOTE.—The preceding pages are published as found in the 3d or Revisers' Edition of the R. S., it being the last authorized edition published. The law of 1847 (chap. 455) amends many sections of the Statutes, to which a reference is made in each instance; and the law is also published herewith in full.

Chap. 225.

GENERAL LAWS PERTAINING TO HIGHWAYS, BRIDGES, &c.

AN ACT *relating to the joint liability of commissioners of highways.*

Passed May 25, 1841.

The People of the State of New York, represented in Senate and Assembly, do enact as follows:

§ 1. Whenever any two or more towns shall be liable to make or maintain any bridge or bridges, the same shall be built and maintained at the joint expense of said towns, without reference to town lines. Bridges, how to be built and repaired.

§ 2. For the purpose of building and maintaining such bridges, it shall be lawful for the commissioners of said towns, or of commissioners of either one or more towns respectively, the other or others refusing to act, to enter into joint contracts, and such contracts may be enforced in law or equity, against such commissioners or their representative successors, jointly or severally respectively; and the commissioners of said towns so liable may be proceeded against jointly for any neglect of duty in reference to such bridges. Commissioners to make contracts.

§ 3. If the commissioners of highways of either of such towns, after notice in writing from the commissioners of highways of any other of such towns, shall not within twenty days give their consent in writing to build or repair any such bridge, and shall not within a reasonable time thereafter do the same, it shall be lawful for the commissioners so giving such notice to make or repair such bridge, and then to maintain a suit at law in their official capacity, against said commissioners so neglecting or refusing to join in such making or repairing, and in such suit the plaintiff or plaintiffs shall be entitled to recover so much from the defendant or defendants respectively representing said other towns as the town or towns Refusal to repair bridges.

would be liable to contribute to the same, together with costs of suit and interest, without proving any contract; and in an action in pursuance of the act hereby amended, to recover the expense of building or repairs, it shall not be necessary to entitle such commissioner or commissioners to recover on the trial of the above action to prove that the defendants, or their predecessors in office, were at the time of the service of the notice above mentioned, in the possession of funds belonging to the town which he or they represent, sufficient to make such repairs, nor shall the want of funds be any defense to the said action; and it shall be the duty of the board of supervisors of the county in which such towns are located, to levy the amount of any judgment so obtained, with costs and interest, on the taxable property of any town, against the commissioner or commissioners of which such judgment has been so obtained, but the commissioner or commissioners of such town shall not be personally liable for such judgment.

Judgment recovered to be a charge on town.

§ 4. Any judgment recovered against the commissioners of highways in their official capacity under the provisions of this act, shall be a charge on said town, and collected in the same manner as other town charges, except in cases where the court before which the judgment shall be recovered shall certify that the neglect or refusal of said commissioners was willful and malicious, in which case said commissioners shall be personally liable for such judgment, and the same may be enforced against them in the same manner as against individuals.

NOTE.—This Act is published here as amended by chap. 383, Laws of 1857.

Chap. 455.

AN ACT *to amend an act entitled " An act to reduce the number of town officers, and town and county expenses, and to prevent abuses in auditing town and county accounts," passed May* 10, 1845.

Passed December 14, 1847, "three-fifths being present."

The People of the State of New York represented in Senate and Assembly, do enact as follows :

§ 1. The second section of the act entitled " An act to reduce the number of town officers, and town and county expenses, and to prevent abuses in auditing town and county accounts," passed May 10th, 1845, is hereby amended so as to read as follows:

The electors to determine whether one or three highway commissioners shall be chosen.

§ 2. The electors of each town shall have power at their annual town meeting, to determine by resolution whether there shall be chosen one or three highway commissioners, and the number so determined upon shall be balloted for and chosen; and if only one shall be determined upon and chosen, he shall possess all the powers and discharge all the duties of commissioners of highways as prescribed by law, and shall hold his office for one year. And whenever three commissioners shall be chosen in any town they shall be divided by lot by the canvassers, upon the result of the canvass, into three classes, to be numbered one, two and three; the term of office of the first class shall be one year, of the second, two, and of the third, three; and one commissioner only shall thereafter annually be elected in such town, who shall hold his office for three years, and until a successor shall be duly elected or appointed; but in case any commissioner shall be elected to fill a vacancy, he shall hold the office only for the unexpired term which shall have

Their powers and term of office.

become vacant; and if two vacancies shall be required to be filled, the canvassers shall, after the canvass, determine by lot as aforesaid, the terms they shall respectively hold. And when any vacancy shall happen by death, removal, resignation, neglect to qualify, or refusal to serve, it shall be supplied until the next succeeding annual town meeting by an appointment in writing, under the hands of any three justices of the peace, or two justices and the supervisor of the town; and every commissioner of highways shall be authorized to administer oaths to any witnesses or juries, in proceedings which may be had by or before them; and whenever any town shall have determined upon having three commissioners, and shall desire to return two or have but one, such town shall have the power so to do by a resolution taken at an annual town meeting, and when such resolution shall have been adopted, no other commissioner shall be elected or appointed, until the term or terms of those in office at the time of adopting such resolution shall expire or become vacant; and they shall have power to act until their terms shall severally become vacant or expire, as fully as if the three continued in office.

§ 2. The fifth section of the said act is hereby amended so as to read as follows:

Damages, how to be assessed.

§ 5. Wherever any damages are now allowed to be assessed by law when any road or highway shall be laid out, altered or discontinued in whole or in part, such damages shall be assessed by not less than three commissioners to be appointed by the county court of the county in which such road or highway shall be, on the application of the commissioner or commissioners of the town; and the commissioners so appointed shall take the oath of office prescribed by the constitution, and shall proceed, on receiving at least six days' notice of the time and place, to meet the highway commissioners and take a view of the premises, hear the parties and such witnesses as may be offered, before them; and they shall all meet and act, and shall assess all damages which may be required to be assessed on the same highway, and shall be authorized to administer oaths to witnesses which may be produced before them under this section, and when they shall all have met and acted, the assessment agreed to by a majority of them, shall be valid; and when so made shall be delivered to a commissioner of highways of the town, who, within ten days after receiving it, shall file it in the town clerk's office.

Provision in case persons conceive themselves aggrieved.

§ 3. Any person conceiving himself aggrieved, or the commissioner or commissioners on the part of the town feeling dissatisfied by any such assessment, may, within twenty days after the filing thereof as aforesaid, signify the same by notice in writing, and serving the same on the town clerk and on the opposite party, that is, the persons for whom the assessments were made or the commissioner or commissioners of highways as the case may be, asking for a jury to reassess the damages and specifying a time not less than ten nor more than twenty days from the time of filing said assessment, when such jury will be drawn at the clerk's office of an adjoining town of the same county by the town clerk thereof; which notice shall be served upon said opposite party within three days after service upon the town clerk as aforesaid, and may be served personally or by being left at the dwelling house of the party with some person in charge thereof, or if there be no such person, or the house be closed, then by fixing the same upon the outer door of said dwelling house.

Names of jurors to be put in box and drawn.

§ 4. At the time and place mentioned in the preceding section, the town clerk of such adjoining town, having received three days' previous notice that such jury is to be drawn, from the person or party asking a reassessment, shall deposit in a box the names of all such persons then residents of his town, whose names are on the last list filed in said town clerk's office of those selected and returned as jurors, pursuant to article second, title four, chapter seventh, part third of the Revised Statutes, who are not interested in the lands through which such road shall be located, nor of kin to either or any of the parties, and shall draw therefrom the names of twelve persons, and shall make a certificate of such names and the purposes for which they were drawn, and shall deliver the same to the party first asking for the reassessment.

Jury, when to be summoned.

§ 5. The party receiving such certificate shall, within twenty-four hours thereafter, deliver the same to a justice of the peace of the town wherein the damages are to be assessed; and it shall be the duty of such justice forthwith to issue a summons to one of the constables of his town, directing him to summon the persons named in said certificate, and shall specify a time and place in said summons at which the persons to be summoned shall meet, but no meeting of such persons shall be had within twenty days from the time of filing the assessment of damages in the town clerk's office by the commissioner or commissioners of highways.

Twelve jurors to be drawn to reassess damages.

§ 6. Upon such persons appearing at the time and place mentioned in the summons, the justice who issued the summons shall draw by lot six of the persons attending to serve as a jury, and the first six persons drawn who shall be free from all legal exceptions, shall be the jury to reassess all the damages required to be assessed upon the same highway; and the said jury shall be sworn by the said justice well and truly to determine and reassess such damages as shall be submitted to their consideration, and shall take a view of the premises, hear the parties and such witnesses as may be offered by the parties, and sworn by said justice before them and shall render their verdict in writing under their hands, which shall be certified by said justice and be delivered to the commissioners of highways of the town, and the same shall be final.

§ 7. In all cases of assessments of damages under the provisions of this act by commissioners appointed by a county court, the costs thereof shall be paid by the town in which the damages shall be assessed, and in cases of reassessments of damages by a jury on the application of the commissioners of highways of any town, and the first assessment shall be reduced thereby, the costs of such assessment shall be paid by the party claiming the damages, otherwise by the town; and in case a reassessment of damages shall be had on the application of the party for whom the damages were assessed, and such party shall fail to increase the same, he shall pay the costs thereof, but when such damages shall be increased by the jury the costs shall be paid by the town; and when applications shall be made by two or more persons for the reassessment of damages by a jury, such jury shall be obtained in conformity with the terms of the notice first served upon the clerk of the town in which the damages are to be assessed; and all persons who may be liable for costs under this section shall be liable in proportion to the amount of damages respectively assessed to them by the first assessment, and may be recovered in an action of assumpsit at the suit of any person or persons entitled to the same before a justice of the peace. Costs by whom to be paid.

§ 8. Any person who shall conceive himself aggrieved by any determination of the commissioners of highways, either in laying out, altering or discontinuing any road, or in refusing to lay out, alter or discontinue any road, may at any time within sixty days after such determination shall have been filed in the office of the town clerk, appeal to the county judge of the county in the same manner as appeals were heretofore allowed to be brought to three judges under title first, article fourth, chapter sixteenth, part first of the Revised Statutes; and when any appeal shall be brought under this section, the said judge, or in case of his residence in the town, or of his interest in the lands through which the road shall be laid Right of appeal.

out, or in case he is of kin to any of the persons interested in said lands, or in case of his disability for any cause, then one of the justices of the sessions shall, after the expiration of the said sixty days, appoint in writing three disinterested freeholders who shall not have been named by the parties interested in the appeal, and who shall be residents of the county but not of the town wherein the road shall be located, as referees to hear and determine all the appeals that may have been brought within the said sixty days, and shall notify them of their appointment, and deliver to them all papers pertaining to the matters referred to them. Upon receiving notice of appointment the said referees shall possess all the powers and discharge all the duties heretofore possessed and discharged by the three judges, and give the same notices heretofore required to be given under title first, article four, chapter six, part one, aforesaid, and before proceeding to hear the appeal or appeals they shall be sworn by some officer authorized to take affidavits to be read in courts of record, faithfully to hear and determine the matters referred to them.

Pay of referees.

§ 9. Every referee appointed under the preceding section shall be entitled to receive two dollars for every day employed in the hearing and decision of such appeal or appeals, to be paid by the party appealing where the determination of the commissioners shall be confirmed, but where it is reversed, to be a charge upon the county; and when said referees shall make any decision, laying out, altering, or discontinuing any road in whole or in part, it shall be the duty of the commissioners of highways of the town to carry out such decision in the same manner as required in cases of final determinations of appeals as provided by the thirteenth section of the act hereby amended, and such decision shall remain unaltered for the term of four years from the time the same shall have been filed in the office of the town clerk.

§ 10. In all cases of assessments of damages for laying out or altering any private road, the commissioners of highways of the town where such road may be situated, shall serve a notice in the same manner as prescribed in the third section of this act, upon the town clerk of the town, and upon the persons interested in said road, specifying a time when a jury of the town will be drawn by the said town clerk at his office, which time shall be within not less than six nor more than ten days from the time of service of such notice. At the time mentioned in said notice, said clerk shall proceed to draw in the same manner from the jury list last filed, the same number of names as provided for juries in cases of reassessments, under the fourth section of this act; and the names so drawn shall be certified to, and the jurors summoned, drawn and sworn, and all proceedings shall be had, and the verdict shall be made, certified to, and be filed as is provided in cases of reassessments of damages as aforesaid; and the same jury shall assess all damages required to be assessed for the same road. [*This section is repealed by* § 17, *chap*. 174, *Laws of* 1853.]

Provision in case of damages for laying out private roads.

§ 11. All damages which may be assessed for laying out or altering any private road, together with the costs of such assessment, shall be paid by the person or persons applying for such road.

Damages and costs, by whom to be paid.

§ 12. All highway appeals which were pending before three judges of county courts on the first Monday of July last, and now remain undetermined, shall be deemed as still pending, and the judges before whom such appeals were pending, shall have full power, and it shall be their duty to proceed and determine such appeals in the same manner and with like effect as if their terms of office had not expired; and whenever any determination has been made by any commissioners of highways since the said first Monday of July, refusing to lay out, alter, or discontinue any highway, any party conceiving himself aggrieved, may appeal therefrom at any time within sixty

Provision respecting appeals pending on the first Monday of July, 1847.

days after the passage of this act in the manner provided by this act.

§ 13. The twenty-sixth section of said act, is hereby amended so as to read as follows:

Fees of officers for criminal proceedings, by whom to be paid.

§ 26. All fees and accounts of magistrates and other officers for criminal proceedings, including cases of vagrancy, shall be paid by the several towns or cities wherein the offense shall have been committed, and all accounts rendered for such proceedings shall state where such offense was committed, and the board of supervisors shall assess such fees and accounts upon the several towns or cities designated by such accounts; but when any person shall be bound over to the oyer and terminer, or court of sessions, or committed to jail to await a trial in either of said courts, the costs of the proceedings had before the single magistrate, shall be chargeable upon the towns or cities as aforesaid, and the costs of the proceedings had after the person shall have been so bound over or committed, shall be chargeable to the county; but nothing herein contained shall apply to cases of felonies, nor where the proceedings or trial for the offense shall be had before any court of oyer and terminer or court of sessions of the county, and the fines imposed and collected in any such cases, shall be credited to said towns or cities respectively. And whenever any criminal warrant or process shall be issued by any magistrate residing out of the town or city wherein the offense shall have been committed, it shall authorize the officer executing the same, to carry the person charged with an offense under this act, before any magistrate resident and being in the town or city wherein such offense shall have been committed, to be proceeded against according to the provisions of the fifteenth section of this act; but the magistrate issuing such warrant or process, shall not lose any jurisdiction over the trial and proceedings against any such persons by reason of anything herein contained, nor shall such magistrate be allowed any compensation for any further proceedings in any such case beyond issuing such warrant or process.

§ 14. It shall be the duty of clerks of boards of supervisors on the thirty-first day of December, or within ten days previous thereto, in each year, to make out a statement showing: Duty of clerks of supervisors.

1. The amount of compensation audited by the board of supervisors, to the members thereof, severally, within the year, and the items and nature of such compensation as audited:

2. The number of days the board shall have been in session within such year, and the distance traveled by the members respectively, in attending the meeting of the board:

3. Whether any accounts were audited or allowed without being verified according to law, for any member of the board of supervisors, or any other person, and if any, how much, and for what.

And such statement shall be certified by such clerk, and be printed in a newspaper printed in the county, in the manner that the accounts audited by boards of supervisors are now required by law to be printed, within two weeks after said statement shall be so made out, and it shall be the special duty of such clerk to see that the same is so published, and for every intentional neglect so to do, such clerk shall be deemed guilty of a misdemeanor.

§ 15. Section one hundred and twenty-one of article six, title one, chapter sixteen, part one of the Revised Statutes, is hereby amended so as to read as follows:

§ 121. In case the commissioners of highways of any town shall be dissatisfied with the determination of the board of supervisors of their county, touching an allowance for any such bridges, such determination shall, on the application of the commissioners, be reviewed by the court of sessions of the same county, whose order in the Provision respecting allowance made by the supervisors.

premises shall be observed by every such board of supervisors.

Allowances to town collectors.

§ 16. Town collectors shall be entitled to collect five per cent fees for all unpaid taxes, under the thirtieth section of the act hereby amended; and whenever any such collector shall make return to the county treasurer for any unpaid taxes, he shall add to the several sums so returned by him five per cent, which shall go to the credit of the county, and be collected with said unpaid taxes; and such collector shall be entitled to receive from the county treasury and be paid by the treasurer two per cent as fees for all taxes so returned by him.

Courts to charge grand juries respecting fees.

§ 17. It shall be the duty of every court at which a grand jury shall be summoned, to charge such jury specially to inquire into any violations of law by public officers in demanding, charging or receiving fees to which they are not entitled by law.

Secretary of state to publish in pamphlet all the laws relative to highways.

§ 18. It shall be the duty of the secretary of state to cause all the general statute laws of the state which relate to highways and private roads, to be printed in pamphlet form and stitched, together with such forms and instructions as he may deem necessary, and cause a sufficient number of copies thereof to be sent to the several county clerks, to furnish one for each county clerk's office and town clerk's office in the state, and one to each commissioner of highways of the several towns.

Allowance to town clerks.

§ 19. Town clerks shall be allowed the sum of fifty cents for drawing and certifying a jury as provided by this act, and a constable for summoning such jury shall be allowed two dollars, except when the jury shall be taken from the same town wherein the road is located, in which case he shall be allowed only one dollar. And jurors who shall be summoned from an adjoining town, and shall attend but not serve, shall be entitled each to fifty cents, and if they shall serve, then one dollar; if from the same town and shall attend and not serve, twenty-five cents; if they shall serve, then fifty cents each.

§ 20. If for any cause any commissioner or referee appointed under this act shall be prevented from serving, or shall refuse to serve, the court or officer who appointed him shall have power to appoint another to supply his place. Vacancies, how to be supplied.

§ 21. All orders for the appointment of commissioners or referees under this act shall be filed and recorded in the office of the town clerk of the town in which the road shall be located. Orders to be filed in town clerk's office.

§ 22. Section sixty-four of title one, article four, chapter sixteen, part first of the Revised Statutes, is hereby amended so as to read as follows: Damages, how to be ascertained in certain cases.

§ 64. The damages sustained by reason of the laying out and opening such road may be ascertained by the agreement of the owner and the commissioners of highways, provided such damages do not exceed one hundred dollars, and unless such agreement be made, or the owner of the land shall in writing release all claim to damages, the same shall be assessed in the manner prescribed by law, before such road shall be opened, or worked, or used. Every such agreement or release shall be filed in the town clerk's office, and shall forever preclude such owner from and further claim for such damages.

§ 23. All damages which may be finally assessed or agreed upon by commissioners of highways for the laying out of any road except private roads, shall be laid before the board of supervisors by the supervisor of the town to be audited with the charges of the commissioners, justices, surveyors or other persons or officers employed in making the assessment and for whose services the town shall be liable, and the amount shall be levied and collected in the town in which the road is located, and the money so collected shall be paid to the commissioners of such town, who shall pay to the owner the sum assessed to him, and appropriate the residue to satisfy the charges aforesaid. Damages assessed to be audited by board of supervisors.

Town auditors to make abstracts of accounts for clerk of supervisors.

§ 24. It shall be the duty of boards of town auditors to make annually brief abstracts of the names of all persons who have presented to said board accounts to be audited, the amounts claimed by each of said persons, and the amounts finally audited by them respectively, and shall deliver said abstracts to the clerk of the board of supervisors, and the said clerk shall cause the same to be printed with the statements required to be printed by him by the fourteenth section of this act.

Repeal.

§ 25. The eighth, tenth and eleventh sections of the act hereby amended, and all laws inconsistent with any of the provisions of this act, are hereby repealed.

Counties excepted.

§ 26. Nothing in this act contained shall apply to the counties of Kings, Queens, Suffolk or New York.

§ 27. This act shall take effect immediately.

Chap. 77.

AN ACT *in relation to laying out private roads.*

Passed March 8, 1848.

The people of the State of New York, represented in Senate and Assembly, do enact as follows:

Private roads, how to be laid out.

§ 1. Private roads may be laid out upon application being made to the commissioner or commissioners of highways of any town; and they or he shall proceed in the manner provided in the tenth section of an act passed December 14, 1847, entitled "An act to amend an act, entitled an act to reduce the number of town officers, and town and county expenses, and to prevent abuses in auditing town and county accounts, passed May 10, 1845," to give notice and summon a jury; and all the proceedings required by said tenth section shall be had, and the jury shall determine the necessity of the road, and the amount of damage to be sustained by the opening thereof, and such amount, together with the expenses of the proceedings, shall be paid by the person to be benefited.

§ 2. This act shall take effect immediately.

Chap. 164.

AN ACT *to amend an act entitled "An act to enlarge the powers of boards of supervisors," passed May* 18, 1838.

Passed April 3, 1848.

The People of the State of New York, represented in Senate and Assembly, do enact as follows:

§ 1. The power given to boards of supervisors by subdivision four, of section one, of the act entitled "An act to enlarge the powers of boards of supervisors," passed April 18, 1838, to appoint special commissioners to lay out public highways, shall not be exercised by any board of supervisors, unless the applicant therefor shall prove to such board of supervisors the service of a notice in writing, on a commissioner of highways of each town through and into which any such highway is intended to be laid, at least six days previous to presenting such application, specifying therein the object thereof, and names of persons proposed to be appointed such commissioners.

Provision as to opening public highways.

Chap. 62.

AN ACT *to regulate the construction of roads and streets across railroad tracks.*

Passed March 29, 1853.

The People of the State of New York, represented in Senate and Assembly, do enact as follows:

§ 1. It shall be lawful for the authorities of any city, village or town in this state, who are by law empowered

To lay out streets or highways

across railroad tracks.

to lay out streets and highways, to lay out any street or highway across the track of any railroad now laid or which may hereafter be laid, without compensation to the corporation owning such railroad; but no such street or highway shall be actually opened for use until thirty days after notice of such laying out has been served personally upon the president, vice-president, treasurer or a director of such corporation.

Railroad corporations to take streets or highways across their tracks.

§ 2. It shall be the duty of any railroad corporation, across whose track a street or highway shall be laid out as aforesaid, immediately after the service of said notice, to cause the said street or highway to be taken across their track, as shall be most convenient and useful for public travel, and to cause all necessary embankments, excavation and other work to be done on their road for that purpose; and all the provisions of the act, passed April second, eighteen hundred and fifty, in relation to crossing streets and highways already laid out, by railroads, and in relation to cattle guards and other securities and facilities for crossing such roads, shall apply to streets and highways hereafter laid out.

Penalty for neglect or refusal

§ 3. If any railroad corporation shall neglect or refuse, for thirty days after the service of the notice aforesaid, to cause the necessary work to be done and completed, and improvements made on such streets or highways across their road, they shall forfeit and pay the sum of twenty dollars for every subsequent day's neglect or refusal, to be recovered by the officers laying out such street or highway, to be expended on the same; but the time for doing said work may be extended, not to exceed thirty days, by the county judge of the county in which such street or highway, or any part thereof, may be situated, if, in his opinion, the said work cannot be performed within the time limited by this act.

§ 4. This act shall take effect immediately.

Chap. 174.

AN ACT *in relation to laying out private roads, and discontinuing public highways.*

Passed April 12, 1853.

The People of the State of New York, represented in Senate and Assembly, do enact as follows:

§ 1. An application for a private road shall be made in writing, specifying its width and location, courses and distances, and the names of the owners and occupants of the land through which the road is proposed to be laid out. Applications how made.

§ 2. The commissioner or commissioners, to whom such application shall be made, shall thereupon appoint as early a day as the convenience of the parties interested will allow, when, at his office, a jury will be selected for the purpose of determining upon the necessity of said road, and to assess the damages by reason of the opening thereon. Jury to determine upon necessity of road, and assess damages.

§ 3. Such commissioner or commissioners shall thereupon deliver to the applicant a copy of such application, to which shall be added a notice of the time and place appointed for the selection of such jury, addressed to the owners and occupants of said land. Copy of application and notice to be delivered to applicant.

§ 4. The applicant, on receiving such copy and notice, shall on the same day, or on the next day thereafter, cause such copy and notice to be served upon the persons to whom it is addressed, by delivering to each of them who reside in the same town a copy thereof, or in case of his absence by leaving the same at his dwelling house, and upon such as reside elsewhere by depositing in the post-office a copy thereof to each, addressed to them respectively at their places of residence, and paying the postage thereon, or, in case of infant owners, by like services upon their parent or guardian. Copy and notice to be served upon persons to whom addressed.

List of persons to act as jurors.

§ 5. At such time and place, on due proof of the service of such notice, such commissioners, or in a town where there are more than one, either of them, shall present a list of the names of eighteen persons, residents of said town, and freeholders qualified to act as jurors in courts of record, who are in no wise of kin to such applicant, owner or occupant, or either of them, and not interested in such lands. [*As amended by chap.* 373, *Laws of* 1859.]

Certain number of names may be struck off from list.

§ 6. The owners or occupants of such lands may strike off from such list any number of names not exceeding six; the applicant may in like manner strike off six names or less, and the persons whose names are not stricken off, or if more than six names are left upon the list, then the six persons whose names stand first upon the list shall be the jury for the purpose aforesaid.

Place of meeting of jury.

§ 7. The commissioner shall then appoint some convenient time and place for the jury to meet and be sworn in the premises, and shall summon them accordingly.

Jury to determine and assess damages.

§ 8. If at the time and place last mentioned, all the persons named as such jury shall meet, they shall be sworn well and truly to determine as to the necessity of said road, and to assess the damages by reason of the opening thereof; if one or more of such six persons shall not appear, the commissioner shall summon, of the bystanders or others, so many, free from all legal objection, as will be sufficient to make the number six, who shall be sworn as aforesaid.

Commissioner to swear the jury.

§ 9. Such commissioner is hereby authorized to swear the jury, and to administer any oath necessary to carry this act into effect.

Jury to make up their verdict.

§ 10. The jury shall view the premises, and, after hearing the allegations of the parties, and such witnesses as they may produce, shall proceed to deliberate and make up their verdict; and if they shall determine that the proposed road is necessary, they shall assess the damages to the person or persons through whose land the same is to pass, and deliver their verdict in writing to the commissioners.

§ 11. If the necessity of such private road has been occasioned by the alteration or discontinuance of a public highway running through the lands belonging to the same person or persons through whose lands the private road is proposed to be opened, the jury shall take into calculation the value of the road so discontinued, and the benefit resulting to such person or persons by reason of such discontinuance, and shall deduct the same from the damages assessed for the opening and laying out of such private road.

Jury to take into calculation value of road discontinued.

§ 12. The commissioner shall annex to such verdict the application mentioned in the first section of this act, and hand the same to the town clerk, who shall file the same, and the commissioner or commissioners shall lay out and make a record of said road, as described in the petition of the applicant.

Application to be annexed to verdict, and same to be handed town clerk.

§ 13. In case any accident shall prevent any of the proceedings required by this act to be done on the day assigned, the proceedings may be adjourned to some other day, and the commissioner shall publicly announce such adjournment.

Proceedings may be adjourned.

§ 14. The damages assessed by the jury shall be paid by the party, for whose benefit the road is laid, before the said road shall be opened or used. But in case the assessors of said town shall certify that the necessity of such private road was occasioned by the alteration or discontinuance of a public highway, such damages shall be paid by said town and refunded to the applicant.

Damages to be paid before opening road.

§ 15. Whenever any public highway, or any part thereof, by reason of alterations made therein, or by the opening of a new road, or in any other way, shall be abandoned by the public, and is no longer used as a public road, the commissioners or commissioner of highways shall file in the town clerk's office of the town a descrip-

Description of any road abandoned.

tion in writing, signed by them or him, of the road so abandoned, and the same shall thereupon be discontinued.

Roads along division lines, and proceedings regarding same.

§ 16. Whenever a public or private road shall be laid along the division line between the lands of two or more persons, and wholly upon one side of said line, and the lands upon both sides of said division line shall be cultivated or improved; then, and in that case, the person owning or occupying the lands joining said road shall be paid for building and maintaining such additional fence as he may be required to build, or maintain, by reason of the laying out and opening said road; which said damages shall be ascertained and determined in the same manner that other damages are now ascertained and determined in the laying highways or private roads.

Repeal.

§ 17. The act entitled "An act in relation to laying out private roads," passed March eighth, eighteen hundred and forty-eight, and the tenth section of the act entitled an act to amend an act entitled "An act to reduce the number of town officers and town and county expenses, and to prevent abuses in auditing town and county accounts," passed May tenth, eighteen hundred and forty-five—passed December fourteenth, eighteen hundred and forty-seven—are hereby repealed.

§ 18. This act shall take effect immediately.

Chap. 265.

AN ACT *to amend the Revised Statutes in relation to laying out of public roads, and of public roads, and of the alteration thereof in the town of Greenburgh.*

Passed April 15, 1854; three-fifths being present.

The People of the State of New York, represented in Senate and Assembly, do enact as follows:

§ 1. Section sixty-four, article fourth of title one, chapter sixteen of the first part of the Revised Statutes, in its application to the town of Greenburgh, is so amended as to read as follows:

In all cases of the alteration of any road, or the laying out of any new road in the said town, the person applying for the same shall serve a notice on the town clerk of the town, and on a justice of the peace and one of the assessors thereof, asking for a jury to certify to the necessity of the same, and specifying a time not less than ten nor more than twenty days from the time of serving such notice, when such jury will be drawn at the clerk's office of the town, by the town clerk thereof. Roads. Notice to town clerk.

§ 2. At the time and place mentioned, the town clerk of such town having received such notice that such jury is to be drawn, shall, in the presence of a justice of the peace or one of the assessors of the town, deposit in a box the names of all persons then residents of his town whose names are on the list filed in said town clerk's office of those selected and returned as jurors, pursuant to article second, title four, chapter seventh, part third of the Revised Statutes, who are not interested in the lands through which such road is to pass, located, nor of kin to the owner thereof, and shall publicly, in the presence of said justice of the peace or assessor, draw therefrom the names of twenty persons, who shall be freeholders of the town, and shall make a certificate of such names and the purposes for which they were drawn, and shall deliver the same to the person asking for the jury. Jurors.

§ 3. The person receiving such certificate shall deliver the same to a justice of the peace of the town wherein the road is to be laid, and it shall be the duty of such justice, forthwith, to issue a summons to one of the constables of his towns, directing him to summon the persons named in said certificate, and shall specify a time and place in said summons, which shall be the time and place mentioned in the notice at which the person to be summoned shall meet, which shall not be less than ten nor more than twenty days from the issuing thereof. Summoning jury.

Their duty. § 4. If twelve or more of the number shall appear, at the time and place specified in the summons, they shall then be sworn by a justice of the peace, or any officer authorized to administer oaths, well and truly to examine and certify in regard to the necessity and propriety of the highway applied for.

Examination of route. § 5. Section sixty-five of the said article, title, chapter and part, in its application as aforesaid, is so amended as to read as follows: They shall then personally examine the route of such highway, and shall hear any reasons that may be offered for or against laying out or altering the same, as the case may be. If twelve or more of the number thereof shall be of the opinion that such highway is necessary and proper, they shall make and subscribe a certificate in writing to that effect, which shall be delivered to the commissioners of highways of the town.

§ 6. This act shall take effect immediately.

Chap. 255.

AN ACT *to enlarge the powers and duties of commissioners of highways.*

Passed April 10, 1855.

The People of the State of New York, represented in Senate and Assembly, do enact as follows:

§ 1. The commissioner or commissioners of highways in each of the towns of this state, are hereby empowered to bring any action against any railroad corporation, that may be necessary or proper to sustain the rights of the pnblic in and to any highway in such town, and to enforce the performance of any duty enjoined upon any railroad corporation in relation to any highway in the town of which they are commissioners, and to maintain an action for damages or expenses which any town may sustain or may have sustained, or may be put to or may have been put to, in consequence of any act or omission of any such corporation in violation of any law in relation to such highway.

§ 2. Nothing in this act shall be construed as in any manner impairing the right of any person or officer to bring any action now authorized by law.

§ 3. This act shall take effect immediately.

Chap. 491.

AN ACT *to provide for the assessment of damages upon the laying out of public highways through uninclosed, unimproved and uncultivated lands.*

Passed April 15, 1857.

The People of the State of New York, represented in Senate and Assembly, do enact as follows:

§ 1. When a highway shall hereafter be laid out through uninclosed, unimproved and uncultivated lands, the damages shall be assessed in the same manner as if the same were laid out through inclosed, improved and cultivated lands. [*As amended by chap.* 51, *Laws of* 1858.] Assessment of damages.

§ 2. All acts or parts of acts inconsistent with this act, are hereby repealed.

Chap. 615.

AN ACT *to allow the several towns in this state to raise an increased amount of money for the support of roads and bridges, and to provide for increased compensation of commissioners of highways and other town officers.*

Passed April 15, 1857; three-fifths being present.

The People of the State of New York, represented in Senate and Assembly, do enact as follows:

§ 1. Whenever the commissioners of highways of any town in this state shall be of the opinion that the sum now provided by law will be insufficient to pay the Duty of commissioners of highways.

expenses actually necessary for the improvement of roads and bridges, and to pay any balance that may be due for such improvement, it shall be lawful for such commissioners to apply in open town meeting for a vote authorizing such additional sum, to be raised as they may deem necessary for the purposes aforesaid, not exceeding seven hundred and fifty dollars in addition to the sum now allowed by law. The same notice shall be given by the commissioners of their intention to apply for the raising of such additional sum as is now required by law for the raising of money for roads and bridges above the amount of two hundred and fifty dollars.

Compensation.

§ 2. The commissioner of highways in any town in this state where there is but one such officer, shall be allowed the sum of two dollars per day for each day actually and necessarily spent in the discharge of his official duties; and in towns where there is more than one commissioner, they shall receive for such official service each the sum of one dollar and fifty cents for each day actually and necessarily spent therein.

§ 3. The following town officers shall be entitled to compensation at the following rates for each day actually and necessarily devoted by them to the service of the town in the duties of their respective offices:

1. The supervisor (except when attending the board of supervisors), town clerks, assessors, justices of the peace, inspectors of elections, clerks of the polls and overseers of the poor, one dollar and fifty cents a day. [*As amended by chap.* 305, *Laws of* 1860.]

§ 4. All provisions of law inconsistent with this act are hereby repealed.

§ 5. This act shall take effect immediately.

Chap. 103.

AN ACT *to provide for the speedy construction and repair of roads and bridges where the same shall have been damaged or destroyed.*

Passed April 6th, 1858; three-fifths being present.

The People of the State of New York, represented in Senate and Assembly, do enact as follows:

§ 1. In case any road or roads, bridge or bridges, shall be damaged or destroyed by the elements or otherwise, after any town meeting shall have been held, or when too late to give notice as required by chapter six hundred and fifteen of the Laws of eighteen hundred and fifty-seven, then it shall be lawful for the commissioner or commissioners of highways, by and with the consent of the board of town auditors of the town or towns in which such road or bridge shall be situated, to cause the same to be immediately repaired or rebuilt, and the commissioner or commissioners of highways shall present the proper vouchers for the expense thereof to the town auditors, at their next annual meeting, and the said bills shall be audited by them and collected in the same manner as though the amount had been voted at any town meeting as now required.

Bridges, &c., to be repaired by commissioners of highways.

§ 2. The board of town auditors may be convened in special session by the supervisor, or in his absence the town clerk, upon the written request of any commissioner of highways, and the bills and expenses incurred in the erection or repairs of any such roads or bridges, may then be presented to and audited by such board of town auditors; and the supervisor and town clerk shall issue a certificate, to be subscribed by them, setting forth the amount so audited and allowed, and in whose favor, and the nature of the work done and material furnished; and

Town auditors to audit expenses incurred, &c.

such certificate shall bear interest from its date, and the amount thereof, with interest, shall be levied and collected in the same manner as other town expenses.

What accounts may not be allowed.

§ 3. No account for services rendered or material furnished according to the provisions of this act, shall be allowed by such board unless the same shall be accompanied by the affidavit of the party or parties performing such labor or furnishing such material, nor unless the commissioner or commissioners shall certify that such service has been actually performed, and such material was actually furnished, and that the same was so performed or furnished by the request of said commissioner or commissioners, and such board of auditors may require and take such other proof as they may deem proper to establish any claim for such labor and material, and the value therefor.

§ 4. This act shall apply to any instance coming within its provisions which may have occurred in the year eighteen hundred and fifty-seven, in the county of Tompkins.

§ 5. This act shall take effect immediately.

Chap. 324.

AN ACT *to amend chapter two hundred and eighty-one of the Session Laws of eighteen hundred and thirty-six, entitled "An act to protect sidewalks along highways," passed May* 10, 1836.

Passed April 15, 1854.

The People of the State of New York, represented in Senate and Assembly, do enact as follows:

§ 1. Section one of "An act to protect sidewalks along highways," passed May tenth, eighteen hundred and thirty-six, is so amended as to read as follows:

§ 1. It shall be lawful for any person owning or occupying lands adjoining a highway or road to construct a sidewalk within such highway or road along the line of such land; and when a sidewalk shall be so constructed, every person who shall ride or drive a horse or team upon it shall forfeit the sum of one dollar to the use of such owner or occupant, to be sued for in any court having cognizance thereof.

§ 2. The commissioners of highways of the several towns in this state are hereby authorized to expend a part of the highway tax levied in their road districts upon the sidewalks in said districts, and in planting shade trees on the public greens or squares in said towns, provided the roads are always kept in good repair.

§ 3. This act shall take effect immediately.

Chap. 259.

BRIDGE COMPANIES.

AN ACT *to provide for the incorporation of bridge companies.*

Passed April 11, 1848.

The People of the State of New York, represented in Senate and Assembly, do enact as follows:

§ 1. Any number of persons not less than five, may be formed into a corporation, for the purpose of constructing and owning a bridge across any stream of water, as hereinafter provided, upon complying with the following requirements: **Corporations may be formed.**

1. They shall severally subscribe articles of association, in which shall be set forth the name of the corporation, the number of years the same is to continue, which shall not exceed fifty years; the amount of the capital stock of the corporation, which shall be divided into shares of twenty-five dollars each, the number of directors and their names, who shall manage the concerns of the corporation for the first year, and until others are elected; the location of such bridge, and the plan thereof: **Articles of association to be subscribed.**

Names and shares.

2. Each subscriber to such articles of association shall subscribe thereto his name and place of residence, and the number of shares of stock taken by him in such corporation:

Articles, when and where to be filed.

3. Whenever one-fourth part of the amount of the capital stock, specified in the articles of association, shall have been subscribed, and on complying with the provisions of the next section, such articles may be filed in the office of the state engineer and surveyor, and clerk of the county or counties in which the bridge is built; and thereupon the persons who have subscribed the articles of association as aforesaid, and such other persons as shall become stockholders in such company, and their successors, shall be a body corporate, by the name specified in such articles of association, and shall possess the powers and privileges, and be subject to the provisions of titles three and four of chapter eighteen of the first part of the Revised Statutes, so far as those provisions are consistent with the provisions of this act.

Liability of stockholders.

§ 2. All the stockholders of every company incorporated under this act, shall be severally and individually liable, to an amount equal to the amount of the capital stock held by them respectively, to the creditors of such company, for all the debts contracted by the directors or agents of such company for its use, until the whole amount of the capital stock fixed and limited by such company is paid in, and a certificate thereof filed in the offices aforesaid, and the whole capital stock paid in, shall be one-half thereof within one year, and the other half thereof, within two years from the time of the incorporation of such company, and if not so paid in, such corporation shall be dissolved. If the directors of any corporation formed under this act shall contract debts for the company, exceeding in the aggregate the amount of the capital stock, they shall be personally liable for all the debts of the corporation.

§ 3. Such articles of association shall not be filed as aforesaid, until five per cent on one-fourth the amount of the stock of such company fixed as aforesaid shall have been actually paid in, in good faith, to the directors named in such articles of association, in cash, nor until there shall be indorsed thereon, or annexed thereto, an affidavit made by at least three of the directors named in such articles of association, that the amount of stock required by the first section of this act to be subscribed, has been subscribed, and that five per cent on the amount has been actually paid in as aforesaid. **Amount to be paid before filing articles.**

§ 4. A copy of such articles of association filed in pursuance of this act, with a copy of such affidavit indorsed thereon or annexed thereto, and certified to be a copy by the proper officer, shall, in all courts and places, be presumptive evidence of the facts therein contained. **Copy of articles to be evidence.**

§ 5. The business and property of every such corporation shall be managed and conducted by a board of directors, consisting of not less than five, nor more than nine, who shall be chosen, except those for the first year, at such place within a county in which the bridge of such corporation or some part thereof shall be located, as shall be prescribed by the by-laws thereof. The directors shall give notice of every such election, previous to the holding thereof, by publishing the same once in each week for four successive weeks, in a public newspaper, published in each county in which such bridge or any part thereof, shall be located, and if in any such county no such paper shall be published, such notice shall be published in some county adjoining such last mentioned county. All elections of directors shall be by ballot and by a majority of all votes given thereat; and every stockholder being a citizen of the United States and attending in person or by proxy, shall be entitled to one vote for each share of stock which he shall have owned absolutely, or as executor, administrator or guardian, for thirty days previous to such election. No person shall be a director **Election of directors.** **By ballot.**

unless he shall be a stockholder, owning at least four shares of stock, absolutely in his own right or as executor, administrator or guardian, and entitled to vote at the election at which he shall be chosen, nor unless he shall be a citizen of this state; and a majority of the directors shall, at the time of their election, be residents of the county or counties in which such bridge shall be located. Whenever any vacancy shall happen in the board of directors, it shall be supplied until the next election by the remaining directors. The directors of every such company shall be elected in the same month, in each and every year, and such election after the first, shall be held on the first Tuesday of such month, and the directors chosen at any election shall hold their offices to and including the Tuesday next after that appointed by law for holding the election next succeeding that at which they were chosen. If an election of directors shall not be held on the day prescribed by this act for holding the same, the directors in office on that day shall hold their offices until their successors shall be elected, but after the expiration of their regular term of office as prescribed by this section, they shall be incapable of doing any act, as such directors, except such as may be necessary to give effect to an election of directors. The provisions of the second article of the second title of the eighteenth chapter of the first part of the Revised Statutes, shall apply to every corporation formed under this act, so far as such provisions shall be consistent with the provisions of this act.

Application to be made to board of supervisors for leave to erect bridges.

§ 6. When any bridge corporation shall be desirous of constructing a bridge or any part thereof, in any county, it shall apply to the board of supervisors of such county at the annual or any special meeting thereof, for authority to construct such bridge; of which application, such corporation shall give notice, by publishing the same in at least one public newspaper in such county, or if no newspaper is published therein, then in an adjoining county, once in each week for six weeks successively, previous to the time of presenting such application to such board, specifying such time and the location of such proposed bridge. If the place of the location of such bridge shall be situated in more than one county, such application shall be made to the board of supervisors of every such county. Such application shall also specify the length and breadth of such bridge; and the notice of such application shall set forth all the particulars required to be specified in such application. Upon the hearing of the said application, all persons residing in such county or interested in such application, may appear and be heard in respect thereto. Such board may take testimony in respect to such application, or may authorize it to be taken by a county judge or justice of the peace of such county; and it may adjourn the hearing from time to time. A copy of the articles of association of such corporation certified by the state engineer and surveyor, or by the clerk where such articles are filed, shall be attached to and filed with such application. No such corporation shall be authorized to bridge any stream, in any manner that will prevent or endanger the passage of any raft of forty-five feet in width, or any ark where the same is navigated by rafts or arks. [*See Sec.* 1, *chap.* 372, *Laws of* 1852, *published immediately following this chapter.*]

Assent of the board, &c, to be recorded, &c.

§ 7. If after hearing such application such board shall be of opinion that the public interests will be promoted by the construction of such bridge on the proposed site, it may, if a majority of all the members elected to such board, shall assent thereto, by an order to be entered in its minutes, authorize such company to construct such bridge, as shall have been specified in the application which shall be particularly described in such order. Such corporation shall cause a copy of such order certified by the clerk of such board, with a copy of such application, to be recorded in the clerk's office of such county, before it shall proceed to do any act by virtue thereof; and such board shall cause such application when it shall have finally acted on the same, to be filed at the expense of the corporation, with all the other papers relating thereto, or to the proceedings of said board thereon, in the office of the clerk of the county in which it shall have been made. Any corporation formed under this act, may use in such manner as such board shall prescribe, so much of any public highway on either side of any stream, as may be necessary for the construction and maintenance of such bridge and toll houses.

Bridges over navigable streams by rafts.

§ 8. In case any bridge shall be constructed under the provisions of this act, over any stream navigable by rafts, it shall be the duty of the corporation constructing such bridge, at all times to keep the channel of said stream both above and below said bridge, free and clear from all deposits in any wise prejudicial to the navigation thereof, which may be formed or occasioned by the erection of such bridge.

Penalty for delaying rafts.

§ 9. Any corporation organized under the provisions of this act, which shall construct any bridge over any stream, navigable by rafts as herein before provided, shall be liable to pay all persons who may be unnecessarily or unreasonably hindered or delayed in passing such bridge, all damages which they shall sustain thereby, to be recovered with costs of suit.

Bridges, how to be constructed.

§ 10. Every bridge constructed by virtue of this act, shall be built with a good and substantial railing or siding, at least four and a half feet high. Whenever such bridge shall be completed, and a certificate signed by the county judge of the county in which such bridge is situated, or if such bridge shall be located in more than one county, by the county judge of each of such counties, and such certificate filed in the office of the clerk of such county or of each of said counties, if such bridge shall be located in more than one county, that such bridge is constructed and completed in a manner safe and convenient for the public use, the directors may erect a toll gate at such bridge, and demand and receive such sum as shall be from time to time prescribed by the supervisors of the county or counties where the bridge is located.

§ 11. No tolls shall be collected for crossing any bridge constructed by any corporation formed under this act, from any person going to or from public worship, or to or from a funeral, or to or from school, or to or from a town meeting or election, at which he is entitled to vote, for the purpose of giving such vote, and returning therefrom; or to or from a military parade which he is by law required to attend, or to or from any court which he shall be required to attend as a juror, or witness or to or from his legally required work upon any public highway. Exemptions from toll.

§ 12. The directors of any incorporation formed under this act, may require payment from the stockholders of the sums subscribed to the capital stock, at such times and in such proportions and on such conditions as they shall see fit, under the penalty of the forfeiture of their stock and all previous payments thereon; and they shall give notice of the payments thus required, and of the place and time when and where the same are to be made, at least thirty days previous to the time fixed for the payment of the same, for the time and in the manner herein before prescribed for giving notice of the election of directors, and by sending such notice to such stockholder by mail, directed to him at his usual place of residence. Calls on stockholders.

§ 13. The shares of any corporation formed under this act shall be deemed personal property, and may be transferred in such manner as shall be prescribed by the by-laws of such corporation; and the directors of every such corporation may, at any time, with the consent of a majority in amount of the stockholders in such corporation, provide for such increase of the capital stock thereof as may be necessary for the completion or reconstruction of such bridge, and the certificate of the amount of any such increase, within thirty days thereafter, shall be filed Transfers of shares.

in the offices of the state engineer and surveyor, and the clerk or clerks of the county or counties in which such bridge is located, which certificate shall be authenticated by the signatures and oaths of a majority of said directors.

Taxation.

§ 14. So much of any such bridge or tollhouses constructed by virtue of this act, as shall be within any town, city, or village, shall be liable to taxation in such town, city or village, as real estate

Companies, when to cease to be bodies corporate.

§ 15. Every company incorporated under this act, shall cease to be a body corporate:

1. If within two years from the filing of their articles of association, they shall not have commenced the construction of their bridges and actually expended thereon at least ten per cent of the capital stock of such company; or,

2. If within five years from the filing of such articles of association such bridge shall not be completed according to the provisions of this act; or,

3. If, in case the bridge of such company shall be destroyed it shall not be reconstructed within three years thereafter.

Annual reports to be made to state engineer and surveyor.

§ 16. It shall be the duty of the president and secretary of every corporation formed under this act, to report annually to the state engineer and surveyor, and the county clerk where the papers are filed, under oath, the costs of their bridge, the amount of all money expended, the amount of their capital stock, and how much paid in, and how much actually expended, the amount received during the year for tolls, and from all other sources, stating each separately, the amount of dividends made, and the amount of indebtedness of such company, specifying the object for which the indebtedness accrued; and such other particulars in respect to the business affairs of such corporation, as the said state engineer and surveyor, or the legislature, or either branch thereof require to be so reported.

§ 17. When any bridge may be in process of construction by private subscriptions at the time of the passage of this act, the snbscribers may organize into a corporation pursuant to the provisions of this act, with the same power and privileges as if such bridge had not been so commenced. Private bridges.

§ 18. All companies formed under this act shall at all times be subject to visitation and examination by an officer or agent, in pursuance of law, or by the legislature, or by a committee, appointed by either house thereof; and the courts of this state shall have the same jurisdiction over such corporations and their officers as over those created by special acts. Companies subject to visitation.

§ 19. Every report required to be made by the sixteenth section of this act, shall be made in the month of January in each year, and shall show, in respect to the particulars required therein to be set forth, the affairs and business of the corporation, making the same at the close of the year, ending on the thirty-first day of December, next preceding the time of making the same, and shall be published in the nearest newspaper four weeks, and every corporation formed under this act, which shall neglect to make such report as thereby required, shall forfeit to the people of this state for every such neglect the sum of two hundred dollars, and for every week such corporation shall neglect to make such report after the expiration of the time, within which it is required as aforesaid to make the same, it shall forfeit as aforesaid the further sum of fifty dollars. The state engineer and surveyor shall report to the attorney-general every such forfeiture, by whom the same shall be sued for and recovered with the costs in the name of the people; and the certificate of the said state engineer and surveyor of any such neglect shall be presumptive evidence thereof, and if any such river, water-course or lake, now so navigable, shall hereafter be rendered navigable up stream by vessels or steam- Reports to be made in the month of January and penalty for neglect.

boats, power to require such bridge to be altered or removed is reserved to the legislature.

Saving clause.

§ 20. Nothing in this act shall be construed so as to authorize the bridging of any river or water-course where the tide ebbs and flows, or any water used for a harbor, any lake, river or water, which is navigable by sail vessels or steamboats, nor the construction of any bridge within the limits prescribed by any existing law for the erection or maintenance of any other bridge.

Privileges of existing bridge companies.

§ 21. Any existing corporation having for its object the construction and maintenance of any bridge whose charter shall expire, may be continued as such corporation by complying with the provisions of this act, so far as the same are applicable to them, with the consent of the supervisors of the county or counties in which their bridge is located, to be obtained on application to them as herein before provided.

§ 22. This act shall take effect immediately.

Chap. 372.

AN ACT *to amend "An act to provide for the incorporation of bridge companies," passed April eleventh, one thousand eight hundred and forty-eight.*

Passed April 16, 1852.

The People of the State of New York, represented in Senate and Assembly, do enact as follows:

§ 1. Whereas, doubts have arisen in regard to the construction of the sixth section of the "Act to provide for the incorporation of bridge companies," passed April eleventh, one thousand eight hundred and forty-eight, which requires that corporations desiring to erect bridges under the act aforesaid, shall give notice of their intention to apply to the supervisors for authority to erect such bridge, and several incorporations have given such notice before being actually incorporated as required in the first and third sections of said act, but where so incorporated when the applications were presented to such boards:

Be it enacted: that the proceedings of such boards of supervisors in granting such authority, shall be considered as valid and effectual as if such notice had been given after such incorporation had been duly formed as aforesaid; provided, such associations were duly incorporated in all respects, at the time their applications were made to the boards of supervisors aforesaid. **Proceedings of supervisors valid.**

Chap. 120.

AN ACT *to amend the act passed in* 1848, *entitled "An act to provide for the incorporation of bridge companies."*

Passed April 1, 1854.

The People of the State of New York, represented in Senate and Assembly, do enact as follows:

§ 1. Every person who shall willfully break, throw down or injure any gate erected on any bridge erected or constructed under and by virtue of the act passed April eleventh, eighteen hundred and forty-eight, entitled "An act to provide for the incorporation of bridge companies," or forcibly or fraudulently pass any such gate thereon, without having first paid the legal toll for crossing said bridge, shall for each offense forfeit to the corporation injured the sum of twenty-five dollars, in addition to the damages resulting from such wrongful act.

§ 2. This act shall take effect immediately.

Chap. 135.

FERRY COMPANIES.

AN ACT *to authorize the formation of corporations for ferry purposes.*

Passed April 9, 1853.

The People of the State of New York, represented in Senate and Assembly, do enact as follows:

Corporation, how formed. § 1. Any three or more persons, desirous of forming a company for the purpose of conducting and managing a ferry, may make, sign and acknowledge, before some officer authorized to take the acknowledgment of deeds, the certificate described in the second section of this act, and cause the same or a copy thereof, executed and acknowledged in like manner, to be filed in the clerk's office of the county or counties in which such ferry shall be or is intended to be established, and in the office of the secretary of state; and thereupon the persons signing such certificate, and such others as may become stockholders of the said company, shall be constituted and shall be a body corporate and politic by the name and for the term expressed in such certificate, and shall possess the powers enumerated in section one of title three, chapter eighteen and part one of the Revised Statutes.

Certificate. § 2. Such certificate shall state the name of the company, the places to and from which the ferry established or to be established shall run, the term, not exceeding fifty years, for which the company is to exist, the amount of the capital stock of such company and the amounts to which it may be increased, the number of shares of which it shall consist, the number of the directors, and the names of those who shall be the directors for the first year.

Affairs of company to be managed by directors. § 3. The affairs of such company shall be managed by a board of directors, not less than three nor more than fifteen in number, who shall be citizens of the United

States, and a majority of whom shall be citizens of this state. Those named in the certificate shall be directors for the first year, and the like number of directors shall be chosen annually thereafter by the stockholders at such time and place as shall be fixed by the by-laws of such company.

Notice of time and place of holding election for directors.

§ 4. Notice of the time and place of holding such election shall be given to the stockholders personally, or by leaving a written or printed notice at their respective dwellings or places of business, or by publication once in each week, for three successive weeks, in one or more of the newspapers published in the county or counties in which such ferry shall be established, if any such papers are there published, and if not, in a newspaper published in an adjoining county; and the election shall be made by such of the stockholders as shall attend for that purpose, either in person or by proxy; all elections shall be by ballot, and each stockholder shall be entitled to as many votes as he owns shares of stock in the said company, and the persons receiving the greatest number of votes shall be directors; and when any vacancy shall happen in the board of directors by death, resignation or otherwise, it shall be filled for the remainder of the year, in such manner as may be provided for by the by-laws of the said company.

Vacancies may be filled.

Company not dissolved for not holding election on day designated.

§ 5. In case it shall happen, at any time, that an election of directors shall not be made on the day designated by the by-laws of the said company, when it ought to have been made, the company for that reason shall not be dissolved; but it shall be lawful, on any other day, to hold an election for directors in such manner as shall be provided for by the by-laws of the company, and all acts of the directors shall be valid and binding until their successors shall be elected.

President and officers

§ 6. There shall be a president of the company, who shall be designated from the number of the directors, and also such subordinate officers as the company by its by-laws may name, who may be appointed, and be required to give such security for the faithful performance of the duties of their office as the company by its by-laws may require.

Eligibility to the office of director.

§ 7. Stockholders only shall be eligible to the office of director; and if a director shall cease to be a stockholder, his office shall be deemed to be vacant.

What shall constitute a vote.

§ 8. Each share of stock shall entitle the holder to one vote on all questions submitted to the stockholders for action.

Subscriptions may be called in by directors.

§ 9. The directors may call in the subscriptions to the stock at such times and in such installments as they may deem proper; and in case any stockholder shall refuse or neglect to make payment according to such call, the directors may require him to make such payment within twenty days thereafter, by serving a notice to that effect upon him personally, or by leaving a copy thereof at his residence or place of business, or by publishing the same in the manner prescribed by the fourth section of this act, upon pain of forfeiting his stock and all previous payments thereon; and in case the payment shall not be made as thus required, such stock, and all the previous payments thereon, shall be forfeited to the company.

Stock personal property, and transferable.

§ 10. The stock of such company shall be deemed personal estate, and shall be transferable in such manner as shall be prescribed by the by-laws of the company.

Certified copy of certificate to be evidence of incorporation.

§ 11. A copy of any certificate of incorporation, filed in pursuance of this act, certified by the county clerk or his deputy to be a true copy thereof and of the whole of such certificate, shall be received in all courts and places as presumptive evidence of the due incorporation of the said company.

Half of capital to be paid in before commencing business.

§ 12. No company formed under this act shall be authorized to commence its business until at least one-half of its capital shall have been actually paid in, nor until affidavits of such payment, sworn to by a majority of the directors, shall have been filed in each of the offices in which the certificate of incorporation is required to be filed by the first section of this act.

§ 13. Any company incorporated under this act shall have power to take by grant, from any authority entitled by the laws of this state to make such grant, or by assignment, the franchise or right to establish and maintain ferries, and to hold and exercise the said franchise or right, and to carry on the business appertaining thereto; but this act shall not be construed to confer any such franchise or right to impair, establish, admit or deny any of the rights of the mayor, aldermen and commonalty of the city of New York, or any other municipal corporation, or of the owner or owners of any legal existing ferry, or the vested rights of any other corporation whatever.

Right to maintain ferries.

§ 14. The president and a majority of the directors, within thirty days after the payments of the last installments of the capital stock so fixed and limited by the company, shall make a certificate stating the amount of the capital stock so fixed and paid in, which certificate shall be signed and sworn to by the president and a majority of the directors; and they shall, within the said thirty days, file the same in each of the offices in which the certificate of incorporation is required to be filed by the first section of this act. The stockholders of any company organized under the provisions of this act shall be jointly and severally individually liable for all debts that may be due and owing to all their laborers, servants and apprentices for services performed for said company; and they shall, in addition to the liabilities hereinabove imposed, be severally individually liable to the creditors of the company in which they are stockholders, to an amount equal to the amount of stock held by them respectively, for all debts and contracts made by such company, until the whole amount of capital stock fixed and limited by such company shall have been paid in, and a certificate thereof shall have been made and recorded as prescribed in this section; and the capital stock so fixed and limited shall

Certificate stating amount of capital stock paid in, &c.

Liability of stockholders.

all be paid in, one-half thereof within one year, and the other half thereof within two years from the incorporation of said company, or such corporation shall be dissolved.

Capital stock may be increased.

§ 15. The capital stock of any such company may be increased from time to time, until it reaches the limit specified in the certificate of incorporation, by a vote of the stockholders represented, not less than a majority of the whole stock of the company, at any annual meeting, or special meeting called for the purpose, and not otherwise; and when any such increase shall have been thus determined upon, a certificate thereof, signed by a majority of the directors, shall, within ten days thereafter, be filed in each of the offices in which the original certificate of incorporation shall have been filed.

Directors to report to stockholders amount of capital paid in, &c.

§ 16. The directors of any such company shall, at each annual meeting of the stockholders, and at every special meeting where directors are to be elected, submit to the stockholders a report, showing the amount of the capital stock of the company actually paid in, the property and effects of the company on hand, the debts due from the company, and the names and places of residence of the stockholders as nearly as the same can be ascertained; and shall also, within ten days thereafter, cause such report, with an affidavit sworn to by a majority of them, to be filed in the offices in which the original certificate of incorporation shall have been filed.

Directors liable for false report.

§ 17. If any certificate or report made by the directors of any such company, in pursuance of the provisions of this act, shall be false in any material representation, all the directors who shall have signed the same, knowing it to be false, shall be jointly and severally liable for all the debts of the company contracted while they were directors thereof.

§ 18. This act shall take effect immediately.

Chap. 210.

GENERAL LAWS IN RELATION TO PLANK AND TURNPIKE ROADS.

AN ACT *to provide for the incorporation of companies to construct plank roads, and of companies to construct turnpike roads.*

Passed May 7, 1847.

The People of the State of New York, represented in Senate and Assembly, do enact as follows:

§ 1. Any number of persons not less than five, may be formed into a corporation, for the purpose of constructing and owning a plank road, or a turnpike road, by complying with the following requirements: Notice shall be given in at least one newspaper, printed in each county through which said road is intended to be constructed, of the time and place or places where books for subscribing to the stock of such road will be opened; and when stock to the amount of at least five hundred dollars for every mile of the road so intended to be built shall be in good faith subscribed, and five per cent paid thereon, as hereinafter required, then the said subscribers, may, upon due and proper notice, elect directors for the said company; and thereupon, they shall severally subscribe articles of association, in which shall be set forth the name of the company, the number of years that the same is to continue, which shall not exceed thirty years from the date of said articles, whether it is a plank road or a turnpike, which the company is formed to construct; the amount of the capital stock of the company; the number of shares of which the said stock shall consist; the number of directors and their names, who shall manage the concerns of the company for the first year, and shall hold their offices until others are elected; the place from and to which the proposed road is to be constructed; and Corporation, how to be created.

each town, city and village into or through which it is intended to pass, and its length, as near as may be. Each subscriber to such articles of association, shall subscribe thereto his name and place of residence, and the number of shares of stock taken by him in said company. The said articles of association may, on complying with the provisions of the next section, be filed in the office of the secretary of state, and thereupon, the persons who have so subscribed, and all persons who shall from time to time, become stockholders in such company, shall be a body corporate, by the name specified in such articles, and shall possess the powers and privileges, and be subject to the provisions contained in titles three and four of chapter eighteen, of the first part of the Revised Statutes.

Articles of association, where to be filed.

§ 2. Such articles of association shall not be filed in the office of the secretary of state, until five per cent on the amount of the stock subscribed thereto, shall have been actually and in good faith paid, in cash, to the directors named in such articles, nor until there is indorsed thereon, or annexed thereto, an affidavit made by at least three of the directors named in such articles that the amount of capital stock required by the first section has been subscribed, and that five per cent on the amount has actually been paid in.

Copy of articles to be evidence.

§ 3. A copy of any articles of association filed in pursuance of this act with a copy of the affidavit aforesaid indorsed thereon or annexed thereto, and certified to be a copy by the secretary of this state or his deputy, shall in all courts and places be presumptive evidence of the incorporation of such company, and of the facts therein stated.

Company to make application to supervisors.

§ 4. Whenever any such company shall be desirous to construct a plank road or turnpike road through any part of any county, it shall make application to the board of supervisors of such county at any meeting thereof legally held, for authority to lay out and construct such road, and to take the real estate necessary for such purpose; and the application shall set forth the route and character of the proposed road as the same shall have been described in the articles of association filed as aforesaid. Public notice of the application shall be given by the company previous to presenting the same to such board by publishing such notice once in each week for six successive weeks in all the public newspapers printed in such county, or in three of such newspapers if more than three are published in such county, which notice shall specify the time when such application will be presented to such board, the character of the proposed road, and each town, city and village in or through which it is proposed to construct the same.

§ 5. If such company shall desire a special meeting of the board of supervisors for hearing the same, any three members of such board may fix the time of such meeting, and a notice thereof shall be served on each of the other supervisors of the county, by delivering the same to him personally or by leaving it at his place of residence at least twenty days before the day appointed for such meeting. The expenses of such special meeting and of notifying the members of such board thereof, shall be paid by such company. Special meeting, how called.

§ 6. Upon the hearing of the said application, all persons residing in such county or owning real estate in any of the towns through which it is proposed to construct such road, may appear and be heard in respect thereto. Such board may take testimony in respect to such application, or may authorize it to be taken by any judicial officer of such county, and it may adjourn the hearing from time to time. Owners of land may be heard.

§ 7. If after hearing such application such board shall be of the opinion that the public interests will be promoted by the construction of such road on the proposed route as shall be described in the application, it may if a majority of all the members elected to such board shall Application when to be assented to by supervisors.

assent thereto, by an order to be entered in its minutes, authorize such company to construct such a road upon the route specified in the application, and to take the real estate necessary to be used for that purpose, a copy of which order certified by the clerk of such board the said company shall cause to be recorded in the clerk's office of such county before it shall proceed to do any act by virtue thereof. [*See Chap.* 360, *Laws of* 1848, *published in this pamphlet.*]

Commissioners to be appointed to lay out road.

§ 8. Whenever any such board shall grant such an application, it shall appoint three disinterested persons who are not the owners of real estate in any town through which such road shall be proposed to be constructed, or in any town adjoining such town, commissioners to lay out such road; the said commissioners after taking the oath prescribed by the constitution shall proceed without unnecessary delay to lay out the route of such road in such manner as in their opinion will best promote the public interest; they shall hear all persons interested who shall apply to them to be heard, they may take testimony in relation thereto, they shall cause an accurate survey and description to be made of such route and of the land necessary to be taken by such company for the construction of such road and the necessary buildings and gates, they shall subscribe such survey and acknowledge its execution as the execution of deeds is required to be acknowledged, in order that they may be recorded, and they shall cause such survey to be recorded in the clerk's office of such county. If such company shall intend to construct its road continuously in or through more than one county, such application shall specify the number of commissioners which the company desire to have appointed to lay out such road, which shall not exceed three for each county, and an equal number of such commissioners shall be appointed by the board of supervisors of each county in or through which it shall be proposed to construct such road, but the whole number of such commissioners shall not be less than three, nor without the consent of such company shall it exceed six, unless the number of counties in or through which it is proposed to construct such road shall exceed that number. And the commissioners so appointed shall lay out the whole of such road, and shall make out a separate survey of so much thereof as lies in each county, which shall be subscribed and acknowledged as aforesaid and recorded in the county clerk's office of such county. Such company shall pay each of the said commissioners two dollars for every day spent by him in the performance of his duties as such commissioner, and his necessary expenses.

Provision respecting orchards, &c.

§ 9. No such road shall be laid out through any orchard to the injury or destruction of fruit trees, or through any garden without the consent of the owner thereof, if such orchard be of the growth of four years or more, or if such garden has been cultivated four years or more before the laying out of such road, nor shall any such road be laid out through any dwelling house or buildings connected therewith, or any yards or inclosures necessary for the use and enjoyment of such dwelling without the consent of the owner, nor shall any such company bridge any stream where the same is navigable by vessels or steamboats, or in any manner that will prevent or endanger the passage of any raft of twenty-five feet in width.

Roadway of turnpike.

§ 10. No plank road shall be made on the roadway of any turnpike company without the consent of such company; and any plank road company formed under this act shall have power to contract with any turnpike company connecting therewith for the purchase of the roadway, or part of the roadway, or the stock of such turnpike company, on such terms as may be mutually agreed upon, and in case the purchase of such stock of such turnpike road company, such stock shall be held by such plank road company for the benefit of the stockholders of such plank road company in proportion to the amount of stock held by each stockholder in such plank road company at the time of such purchase or at any time afterwards. Upon and after the purchase of the whole of the stock of such turnpike road company by such plank road company,the directors of such plank road company for the time being, and their successors, shall be the sole directors of such turnpike road company, and shall manage the affairs thereof pursuant to the charter of such turnpike road company, and shall render an account of the same annually to the stockholders of such plank road company. In case of a dissolution of such plank road company, the stockholders of such plank road company at the time of such dissolution, shall be the stockholders of such turnpike road company in proportion to the amount of stock held by each in said plank road company; and from thenceforward the stock of such turnpike road company shall be deemed divided into shares equal in number to the shares of stock of such late plank road company, and scrip therefor shall be issued accordingly to each of the last stockholders of such plank road company. Whereupon the officers of such turnpike road company shall be the same in number and power as provided for in the charter of such turnpike road company, and shall be chosen by such former stockholders of such plank road company or their assigns, each share of stock as above provided for entitling the holder thereof to one vote. After such purchase of the stock of such turnpike road company, and prior to the dissolution of such plank road company, the assignment of stock in said plank road company shall carry with it its proportional amount of the stock in such turnpike road company, and entitle the holder thereof to his share of the dividends derived from such turnpike road. Whenever a plank road shall be made as provided in this act on or adjoining a route of any turnpike road, the company owning such turnpike road is authorized to abandon that portion of

In case of dissolution.

their road on or adjoining the route of which a plank road is actually constructed and used; but nothing herein contained shall be so construed as to prevent any plank road from crossing any turnpike road, nor any turnpike road from crossing any plank road. [*As amended by chap.* 643, *Laws of* 1857.]

Company may take possession of land.

§ 11. The route so laid out and surveyed by the said commissioners shall be the route of such road, and such company may enter upon, take and hold, subject to the provisions of this act, all such lands as the said survey shall describe as being necessary for the construction of such road and the necessary buildings and gates. But before entering upon any of such lands, the company shall purchase the same of the owners thereof, or shall, pursuant to the provisions of this act, acquire the right to enter upon, take and hold the same.

Provision in cases where land cannot be purchased

§ 12. If any owner of such land shall from any cause be incapable of selling the same, or if such company cannot agree with him for the purchase thereof, or if after diligent inquiry the name or residence of any such owner cannot be ascertained, the company may present to the first judge or county judge of the county in which the lands of such owner lie, a petition setting forth the grounds of the application, a description of the lands in question and the name of the owner if known, and the means that have been taken to ascertain the name and residence of such owner, if his name and residence has not been ascertained, and praying that the compensation and damages of the owner of the lands described in the petition may be ascertained by a jury. Such petition shall be verified by the oaths of at least two of the directors of the company, and if it shall allege that the name or residence of any owner is unknown, it shall be accompanied by affidavits proving to the satisfaction of the said judge that all reasonable efforts have been made by the company to ascertain the name and residence of any owner whose name or residence is unknown.

Case to be submitted to a jury.

§ 13. On receiving such petition, the said judge shall appoint a time for drawing such jury which shall be drawn from the grand jury box of the county by the clerk thereof, at his office. At least fourteen days' notice of the time and place of such drawing shall be served personally upon each owner of lands described in the petition, who shall be known and reside in the county where the lands lie or by leaving the same at his residence, and such notice shall be served on all other owners in the manner aforesaid or by putting the same into the post-office directed to them at their respective places of residence and paying the postage thereon, or by publishing the same once in each week for two successive weeks in a newspaper printed in such county, the first of which publications shall be at least fourteen days before such drawing.

Provision in case of married women, &c.

§ 14. In case any lands described in such petition shall be owned by a married woman, infant, idiot or insane person, or by a non-resident of the state, the said judge shall appoint some competent and suitable person having no interest adverse to such owner to take care of the interests of such owner in respect to the proceedings to ascertain such compensation and damages. And all such notices as are required to be served on any owner residing in such county, shall be served upon the person so appointed in like manner as on such owner; but any person so appointed to take care of the interests of any such non-resident may be superseded by him.

Duty of judge.

§ 15. The said judge shall attend such drawing and shall decide upon any challenge made to any juror drawn by any person interested. Twenty-four competent and disinterested jurors and as many more as the said judge shall direct shall be drawn; the clerk shall make, certify and deliver to the judge and to any party requiring the same a list of them, and the ballots drawn shall be returned to the box. The said judge if he shall deem it necessary, may at any subsequent time direct the drawing of an additional number of jurors, and they shall be drawn, and all proceedings in relation to such drawing shall be had in the manner hereinbefore provided. Before proceeding to draw any such jury the company shall furnish to the said judge proof by affidavit satisfactory to him, of the time and manner of serving and publishing notice of such drawing, which affidavit shall be filed in such clerk's office; and no such jury shall be drawn unless it shall appear to the satisfaction of the said judge that the provisions of this act in respect to giving notice of such drawing have been complied with.

Number of jurors to be drawn.

§ 16. From the jurors so drawn the said judge shall draw as many as he shall deem necessary to secure the attendance of twelve, and he shall issue his precept directed to the sheriff of such county, either of his deputies or any constable of such county to summon the jurors so drawn by the said judge, to attend at the time and place therein specified to ascertain such compensation and damages. And he may from time to time, in case of the absence or inability to serve of any juror directed to be summoned, draw and direct to be summoned as aforesaid, as many as may be necessary in his opinion to secure the attendance of twelve.

Jurors when and how summoned.

§ 17. Every juror named in any such precept, shall, at least four days before the day therein specified for his attendance, be summoned personally, or by leaving at his residence, a notice containing the substance of such precept. The officer serving such precept, shall return it to the said judge, with an affidavit of the manner of serving the same, and of the distance necessarily traveled by him for that purpose; and such officer shall receive for making such service, six cents a mile for the distance so traveled.

Penalty for neglect.

§ 18. Every juror so summoned, who shall neglect or refuse to attend or serve, in pursuance of such summons, shall be liable to the same penalties, as in case of such neglect or refusal of a person duly summoned as a juror in a court of record, and may be excused by the said judge from attending or serving, for reasons for which such juror might be so excused if summoned as a juror in such court. Every juror attending, shall be entitled therefor to one dollar a day, and his reasonable and necessary expenses to be paid by the company.

Witnesses may be subpœnaed.

§ 19. On the application of any party interested, any judge or justice of the peace, may issue a subpœna requiring witnesses to attend before such jury, and such subpœna shall have the same force and effect; and witnesses duly subpœnaed by virtue thereof, and refusing or neglecting to obey the same, shall be subject to the same penalties and liabilities as though such subpœna were issued from a court of record, in a suit pending therein.

Notice to be given to owners of land.

§ 20. The time and place of meeting of the jury, to ascertain such compensation and damages, may be fixed by the said judge, by an order to be made by him at any time after receiving such petition; and notice thereof shall be served on the owners whose lands are described in the petition, as follows: on any owner residing in the county, or within fifteen miles of the lands in question owned by him, personally, or by leaving the same at his residence, at least fourteen days before the time so fixed; on any other owner residing within this state, and whose residence is known, in the manner aforesaid, or by putting the notice into the post office directed to him at his place of residence, and paying the postage thereon; on any owner residing out of the state, and not within fifteen miles of the lands in question, owned by him, by putting the notice in the post office directed and paid as aforesaid, at least forty days before the time so fixed; and on owners whose residence is unknown; by publishing the notice once in each week for six successive weeks, in one of the public newspapers printed in the county.

Duty of jury.

§ 21. The jurors so summoned, shall meet at the time and place fixed by the said judge for that purpose, and shall be sworn by him to diligently inquire and ascertain the compensation and damages which ought justly to be paid for the land described in the petition, or for those of them in respect to which they shall be called upon to inquire, to the owners thereof, and for taking the same for such road, and faithfully to perform their duty as such jurors, according to law.

Duty of judge.

§ 22. The said judge shall attend such jurors, shall administer oaths to witnesses called before them, shall take minutes of the testimony given, and admissions of the parties made before them, shall advise such jury as to the law applicable to any case that may arise, shall receive, certify and return to the county clerk's office, the verdicts agreed upon by them, and while so attending, shall have all the powers possessed by a court of record, when trying issues of fact joined in civil cases.

Damages to be awarded.

§ 23. The jury after hearing the parties, and viewing the lands in question, in each case, shall, by a verdict, ascertain and determine the compensation and damages that ought to be paid to the owner for the land, to be taken by the company, and for taking the same for such road, and also the amount that ought to be paid to him for the time spent, and necessary expenses incurred by him in respect to the proceedings, to ascertain and determine such compensation and damages, of which time and expenses, a bill of items shall be presented to the jury, verified by the oath of the owner or his agent, and such compensation and damages shall be ascertained and determined without any deduction on account of any real or supposed benefit, which the owners of such lands may derive from the construction of such road.

Proof of notice to be produced

§ 24. Such jury shall not proceed to a hearing in any case until the company shall have produced to the said judge, satisfactory proof by affidavit, that the notice of the meeting of the jury has been given in such case, according to the provisions of this act; and such affidavit shall be attached to and filed with the certificate of the verdict in the case. And on any such hearing, no evidence or information shall be given, nor any statement made to the jury, of any proposition by, or negotiation between the parties or their agents, in respect to any such lands, or such compensation or damages, nor shall any such petition contain any such statement or information.

Jury to make certificate.

§ 25. Such jury, finding any such verdict, shall, after agreeing upon the same, make a certificate thereof, and sign and deliver the same to the said judge; and shall embrace therein a particular description of the land, in respect to which it is found. Such certificate may include one or more verdicts, in the discretion of the jury. Every such certificate shall be certified by the judge, to have been made by such jury; and shall be recorded in the records of deeds, in the clerk's office of the county where the lands therein described shall lie, at the expense of the company.

Provision in case of using highways.

§ 26. Whenever it shall become necessary for any such company, to use any part of a public highway for the construction of a plank or turnpike road, the supervisor and commissioners of highways of the town in which such highway is situated, or a majority, if there be more than one such commissioner in such town, may agree with such company upon the compensation and damages to be paid by said company, for taking and using such highway for the purposes aforesaid. Such agreement shall be in writing, and shall be filed and recorded in the town clerk's office of such town. In case such agreement cannot be made, the compensation and damages for taking such highway for such purpose, shall be ascertained in the same manner as the compensation and damages for taking the property of individuals. Such compensation and damages shall be paid to the said commissioners, to be expended by them in improving the highways of such town.

New trial may be applied for.

§ 27. Any party interested in any such verdict, may, within twenty days after being notified of the rendition thereof, apply to the supreme court for a new trial, and it may be granted upon such terms as to the costs of the application and of the first trial, as that court shall deem reasonable. If a new trial shall be granted, a jury shall be drawn therefor, and the same proceedings shall be had as are hereinbefore provided.

§ 28. Within forty days after the rendition of any such verdict, if a new trial shall not be applied for, the company shall pay to the person entitled to receive the same the amount thereof, or shall make a legal tender thereof to him, if he shall refuse to receive the same; and the company may thereupon enter upon the lands in respect to which such verdict was rendered, and take and hold the same to it and its assigns, so long as it shall be used for the purposes of such a road as such company was formed to construct.

Money, when to be paid.

§ 29. If any person entitled to receive the amount of any such verdict be not a resident of this state, or cannot be found therein after diligent search, the company may furnish to the said judge satisfactory proof, by affidavit, of such fact, and he shall thereupon make an order that the amount of such verdict be paid to the treasurer of the county in which the lands lie, in respect to which such verdict was found for the use of such owner, and that notice of such payment shall be given by publishing the same once in each week, for six sucessive weeks in a newspaper published in the county. On satisfactory proof being made to the said judge, by affidavit, within three months from the time of making the last mentioned order, of such payment and publication, he shall make an order authorizing the company to take and hold the land in respect to which such verdict was rendered, in the same manner and with the same effect as if such payment had been made to the owner personally. The affidavit and orders mentioned in this section, and all other affidavits and orders made, and precepts issued in the course of the proceedings under this act, in relation to the ac-

Provision in cases of non-residents.

quisition of the land to be used for such road, shall be filed in the county clerk's office and all such orders shall be recorded by such clerk in the records of deeds, at the expense of the company.

Land to be taken and money deposited in certain cases.

§ 30. If any owner shall apply for a new trial, the company, upon depositing the amount of the verdict sought to be set aside, in such manner as the said judge shall, upon hearing the parties, direct, in trust that the same or so much thereof as the said owner shall be entitled to receive, shall be paid to him on demand, and on giving such security, by bond, as the judge shall approve, for the payment to such owner of any sum which he may be entitled to receive from the company in respect to the land in question, by reason of any verdict or the judgment of any court, for such compensation, damages, costs and expenses, the company may enter upon and use such lands for the purposes of such road, but the title of the owner thereof shall not be divested until the payment or legal tender to him of the whole amount which he shall be entitled to receive from the company for such compensation, damages, costs and expenses; and on such payment or tender being made, the company shall be entitled to take and to hold such lands to it and to its assigns so long as the same shall be used for the purposes of such a road, as such company was formed to construct.

Width of plank roads.

§ 31. Every plank road made by virtue of this act, shall be laid out at least four rods wide, and shall be so constructed as to make, secure and maintain a smooth and permanent road, the track of which shall be made of timber, plank, or other hard material, so that the same shall form a hard and even surface, and be so constructed as to permit carriages and other vehicles conveniently and easily to pass each other, and also so as to permit all carriages to pass on and off where such road is intersected by other roads.

Width of turnpike roads.

§ 32. Every turnpike road that shall be constructed by virtue of this act, shall be laid out at least four rods wide; and shall be bedded with stone, gravel or such other material as may be found on the line thereof, and faced with broken stone or gravel, so as to form a hard and even surface, with good and sufficient ditches on either side wherever the same is practicable. The arch or bed of such road shall be at least eighteen feet wide, and shall be so constructed as to permit carriages and other vehicles conveniently to pass each other, and to pass on and off such turnpike where it may be intersected by other roads.

Inspectors to be appointed

§ 33. In each county of this state, in which there shall be any plank road, or turnpike road, constructed by virtue of this act, there shall be three inspectors of such roads, who shall not be interested in any plank or turnpike road in such county. They shall be appointed by the board of supervisors of the county, and shall hold their offices during the pleasure of such board. Before entering on their duties, such inspectors shall take and subscribe the constitutional oath of office, and file the same in the office of the clerk of the county.

To inspect roads and file certificate.

§ 34. Whenever any such company shall have completed their road, or any five consecutive miles thereof, it may apply to any two of the inspectors to be appointed pursuant to this act, in the county where said road or a part thereof, so completed and to be inspected is located, to inspect the same; or if such inspectors, or a majority of them, are satisfied on inspection, that the road so inspected is made and completed according to the true intent and meaning of this act, they shall grant a certificate to that effect, which shall be filed in the office of the county clerk. The inspectors shall be allowed two dollars per day for their services pursuant to this section, to be paid by the company whose road they inspect.

Rates of toll to be charged.

§ 35. Upon filing as aforesaid such certificate, the company owning any plank road so inspected, may erect one or more toll gates upon their road, but not within three miles of each other, and may demand and receive toll, not exceeding one and a half cents per mile, for any vehicle drawn by two animals, and for any vehicle drawn by more than two animals, one-half cent per mile for every additional animal; for every vehicle drawn by one animal, three-quarters of a cent per mile; for every score of sheep or swine, and for every score of neat cattle, one cent per mile; for every horse and rider, or led horse, half a cent per mile. In no case shall any plank road company charge or receive rates of toll which will enable said company to divide more, nor shall any company divide more than ten per cent per annum on their capital stock actually paid in and invested in their road, after keeping the road in repair, and appropriating not exceeding ten per cent per annum on their capital stock invested as aforesaid, as a fund for the reconstruction of their road when necessary.

Toll gates to be erected.

§ 36. Upon filing such certificate as aforesaid, the company owning any turnpike road so inspected, may erect one or more toll gates upon its road, but not within three miles of each other, and may demand and receive toll not exceeding the following rates: For every vehicle drawn by one animal, three-quarters of a cent a mile; for every vehicle drawn by two animals, one and one-quarter cents a mile; and for every vehicle drawn by more than two animals, one and one-quarter cents a mile, and one-quarter cent additional a mile for every animal more than two; for every score of neat cattle, one cent a mile; for every score of sheep or swine, one-half cent a mile; and in the same proportion for any greater or less number of neat cattle, sheep or swine; for every horse and rider, or led horse, one-half cent a mile: And in no case shall any such turnpike company charge or receive rates of toll which will enable it to divide more than twelve per cent on its capital stock actually paid in, in cash, and invested in its road, after paying the expenses of managing the same, and keeping it in repair.

§ 37. The commissioners of highways of any town in which a toll gate may be located on any such road, or in an adjoining town, whenever they or a majority of them, shall be of the opinion that the location of such gate is unjust to the public interest, by reason of the proximity of diverging roads, or for other reasons, may, on at least fifteen days' written notice to the president or secretary of the said company, apply to the county court of the county in which such gate is located, for an order to alter or change the location of the said gate; the court, on such application, and on hearing the respective parties, and on viewing the premises, if the said court shall deem such view necessary, shall make such order in the matter, as to the said court may seem just and proper; and either party may, within fifteen days thereafter, appeal from such order to the supreme court, on giving such security as said county judge shall require. Such order, unless appealed from, shall be observed by the respective parties, and may be enforced by attachment or otherwise, as the said court shall direct. And if appealed from the decision of the supreme court shall be final in the matter. The said county and supreme court may direct the payment of costs in the premises, as shall be deemed just and equitable.

Location of gates, how changed.

§ 38. The business and property of such company, shall be managed and conducted by a board of directors, consisting of not less than five nor more than nine, who, after the first year, shall be elected at such time and place as shall be directed by the by-laws of such corporation, and public notice shall be given of the time and place of holding such election, not less than twenty days previous thereto, in a newspaper printed in each county in or through which the road of such company is located; the election shall be made by such of the stockholders as shall attend for that purpose, either in person or by proxy; all elections shall be by ballot, and each stockholder shall be entitled to as many votes as he shall own shares of

Board of directors to manage the affairs.

stock, and the persons having the greatest number of votes, shall be directors; whenever any vacancy shall happen in the board of directors, such vacancy shall be filled for the remainder of the year by the remaining directors; the directors shall hold their office for one year and until others are elected in their places; no person shall be a director unless he is a stockholder in the company, and no stockholder shall be permitted to vote at any election for directors, on any stock except such as he has owned for the thirty days next previous to the election.

Calls may be made on stockholders. § 39. The directors of any company incorporated under this act, may require payment of the sums subscribed to the capital stock, at such times, and in such proportions, and on such conditions as they shall see fit, under the penalty of the forfeiture of their stock, and all previous payments thereon; and they shall give notice of the payments thus required, and of the place and time, when and where the same are to be made, at least thirty days previous to the payment of the same, in one newspaper printed in each county, in or through which their road is located, or by sending such notice to such stockholder by mail, directed to him at his usual place of residence.

Shares transferable. § 40. The shares of any company formed under this act, shall be deemed personal property, and may be transferred as shall be prescribed by the by-laws of such company; the directors of every such company, may, at any time, with the consent of a majority, in amount, of the stockholders in such company, provide for such increase of the capital stock of such company as may be necessary to finish the making of a road actually commenced and partly constructed, but the whole capital stock of any company, shall not exceed five thousand dollars per mile, for each mile of road.

Annual report to be § 41. It shall be the duty of the directors of every company formed under this act, to report annually, to the secretary of state, under the oath of any two of such directors, the cost of their road, the amount of all money expended, the amount of their capital stock, and how much paid in, and how much actually expended; the whole amount of tolls or earnings expended on such road; the amount received during the year, for tolls, and from all other sources, stating each separately, the amount of dividends made, and the amount set apart for a reparation fund, and the amount of indebtedness of such company, specifying the object for which the indebtedness accrued. made to secretary of state.

§ 42. Within two weeks after the formation of any company, by virtue of this act, the directors thereof shall designate some place within a county, in which, according to the articles of association of such company, its road, or some part thereof, is to be constructed, as the office of such company; and shall give public notice thereof, by publishing the same in a public newspaper, published in such county, which publication shall be continued once in each week, for three successive weeks, and shall file a copy of such notice in the office of the county clerk of every county, in which any part of such road is constructed or is to be constructed. And if the place of such office shall be changed, like notice of such change shall be published and filed as aforesaid, before it shall take place, in which notice, the time of making the change shall be specified. And every notice, summons, declaration or other paper required by law to be served on such company, may be served by leaving the same at such office with any person having charge thereof, at any time between nine o'clock in the forenoon and noon, and between two and five o'clock in the afternoon of any day except Sunday. Office of compnay to be located.

§ 43. It shall be the duty of the directors of any such company, to cause a book to be kept by the secretary, treasurer or clerk thereof, containing the names of all persons, alphabetically arranged, who are, or shall, within List of stockholders to be recorded, with amount of stock.

six years, have been stockholders of such company, and showing their places of residence, the number of shares of the stock held by them respectively, and the time when they respectively became the holders of such shares; which book shall, from nine o'clock in the forenoon until noon, and from two o'clock in the afternoon until five, on every day except Sunday and the fourth day of July, be open for the inspection of all persons who may desire to examine the same, at the office of such company, and any and every person shall have the right to make extracts from such book; and no transfer of stock shall be valid for any purpose whatever, except to render the person to whom it shall be transferred, liable for the debts of the company, according to the provisions of this act, until it shall have been entered therein, as required by this section, by an entry showing to and from whom transferred. Such book shall be presumptive evidence of the facts therein stated, in favor of the plaintiff in any suit or proceeding against such company, or against any one or more stockholders, or against such company and one or more stockholders jointly. Every officer or agent of any such company, who shall neglect to make any proper entry in such book, or shall refuse or neglect to exhibit the same, or allow the same to be inspected and extracts to be taken therefrom, as provided by this section, shall be deemed guilty of a misdemeanor, and the company shall forfeit and pay to the party injured, a penalty of fifty dollars for every such neglect or refusal, and all the damages resulting therefrom. And every company that shall neglect to keep such a book open for inspection as aforesaid, shall forfeit to the people the sum of fifty dollars for every day it shall so neglect, to be sued for and recovered in the name of the people, by the district attorney of any county in or through which the road of such company shall be constructed, or shall be, according to its articles of association, intended to be constructed, and when so recovered, the amount shall be paid in equal portions to every such county for the use thereof.

§ 44. The stockholders of every company incorporated under this act, shall be liable in their individual capacity for the payment of the debts of such company, for an amount equal to the amount of the stock they severally have subscribed or held in said company over and above such stock, to be recovered of the stockholder who is such when the debt is contracted, or of any subsequent stockholder, and any stockholder who may have paid any demand against such company, either voluntarily, or by compulsion, shall have a right to resort to the rest of the stockholders who were liable to contribution; and the dissolution of any company shall not release or affect the liability of any stockholder which may have been incurred before such dissolution. Liability of stockholders.

§ 45. The debts and liabilities of any company formed under this act shall not exceed in amount, at any one time, fifty per cent of the amount of its capital, actually paid in, and if such debts and liabilities shall at any time exceed such amount, the stockholders who were such, at the time any excess of debts or liabilities shall be created or incurred, shall be jointly and severally, individually liable for such excess, in addition to their other individual liability, as provided in this act. Debts.

§ 46. In any action against any company formed under the provisions of this act, the plaintiff may include as defendants, any one or more of the stockholders of such company, who shall by virtue of the provisions of this act, be claimed to be liable to contribute to the payment of the plaintiff's claim; and if judgment be given against such company, in favor of the plaintiff for his claim or any part thereof and any one or more of the stockholders so made defendants, shall be found to be liable as aforesaid, judgment shall also be given against him or them, and shall show the extent of his or their liabilities individually. The execution upon such judgment shall direct Actions, how brought.

the collection of the sum for which it may be issued, of the property of such company liable to be levied upon by virtue thereof; and in case such property sufficient to satisfy the same, cannot be found in the county of the officer to whom the same shall be directed, that the deficiency, or so much thereof as the stockholders who shall be defendants in such judgment shall be liable to pay, shall be collected of the property of such stockholders respectively. And if in any such action, any one or more of such stockholders shall be found not to be liable for the demand of the plaintiff or any part thereof, judgment shall be given for the stockholders so found not to be liable, but no verdict or judgment in favor of any such stockholder, shall prevent the plaintiff in such action from proceeding therein against the company alone, or against it and such defendants who are stockholders as shall be liable for such demand or some portion thereof. Suits may be brought against one or more stockholders who are claimed to be liable for any debt owing by the company, or any part of such debt, without joining the company in such suit; but no such suit shall be so brought, until judgment on the demand shall have been obtained against the company and execution thereon returned unsatisfied in whole or in part, or the company shall have been dissolved; but it shall not be necessary that such dissolution shall have been declared by any judicial decree, sentence or determination; and in such suit there may be a verdict and judgment in favor of any defendant not liable as aforesaid; but such verdict and jndgment shall not prevent the plaintiff in such suit from proceeding therein against any defendant who shall be liable as aforesaid.

Part of Revised Statutes to apply

§ 47. Sections seven, eight, nine, eleven, twelve, thirteen, fourteen, sixteen, twenty-one, twenty-two, twenty-three, twenty-four, thirty, thirty-five, thirty-six, forty-one, forty-two, forty-four, forty-five, forty-six, forty-seven, forty-eight, forty-nine, fifty, fifty-one and fifty-two, of the first title of the eighteenth chapter of the first part of the Revised Statutes, shall apply to the companies organized by virtue of this act, and all inspectors and other officers named therein, and to all the officers and roads of such companies, so far as the same can be so applied, and are consistent with this act. [*As amended by Chap.* 287, *Laws of* 1847.]

§ 48. So much of any such road and of the toll houses, gates and other appurtenances thereof constructed by virtue of this act, as shall be within any town, city or village shall be liable to taxation in such town, city or village as real estate. Taxation.

§ 49. Every company incorporated under this act shall cease to be a body corporate: Body corporate.

1. If within two years from the filing of their articles of association, they shall not have commenced the construction of their road, and actually expended thereon at least ten per cent of the capital stock of such company; and Time limited

2. If within five years from such filing of the articles of association, such road shall not be completed according to the provisions of this act.

§ 50. All companies formed under this act shall at all times be subject to visitation and examination by the legislature or by a committee appointed by either house thereof, or by any agent or officer in pursuance of law; and the courts of this state shall have the same jurisdiction over such corporations and their officers as over those created by special acts. Companies subject to visitation.

§ 51. The legislature may at any time alter, amend or repeal this act, or may annul or repeal any corporation formed or created under this act. Right to repeal.

§ 52. This act shall take effect immediately.

Chap. 398.

AN ACT *in relation to plank road and turnpike road companies.*

Passed November 24, 1847.

The People of the State of New York, represented in Senate and Assembly, do enact as follows:

Lands, how to be procured by companies.

§ 1. Any company formed under the provisions of chapter two hundred and ten of the Laws of 1847, entitled "An act to provide for the incorporation of companies to construct plank roads, and of companies to construct turnpike roads," may procure by purchase or gift from the owners thereof, any lands necessary for the construction of so much of its contemplated road as shall be intended to be constructed in any county, and may also procure by agreement from the officers named in the twenty-sixth section of the said chapter, the right to take and use any part of any public highway necessary for the construction of so much of such road as shall be intended to be constructed in such county, and when any such company shall have so procured all the lands necessary to be used for the construction of its road in such county, and the right to take and use such parts of the public highways in such county as shall be necessary for that purpose, such company may construct so much of its road as shall be intended to be constructed in such county, without making the application mentioned in the fourth section of the said chapter.

Accurate surveys of road to be made, &c.

§ 2. Before proceeding to construct such part of its road, as provided in the first section of this act, such company shall cause an accurate survey of such part to be made by a practical surveyor, signed by its president and secretary, acknowledged by them as conveyances of real estate are required to be acknowledged in order to be recorded, and recorded in the county clerk's office of such county; and it shall also, before proceeding to construct such part of its road, procure, in the manner provided by the said chapter, from the board of supervisors of every other county, if any there be, in which any portion of its road is intended to be constructed, authority to construct the same through such other county, but in such case, the commissioners appointed to survey and lay out the road of such company, shall not be required to survey and lay out that portion of it intended to be constructed in the county in which such company shall have procured the lands and the right to take and use the public highways necessary for its construction as aforesaid.

§ 3. When any such company by virtue of the provisions of this act shall have procured the lands and the right to take and use the parts of any public highways, necessary to construct its road in any county, and shall have constructed the same without making the application mentioned in the fourth section of the said chapter, it shall possess the same rights, powers and privileges, and be subject to the same duties and liabilities in respect to its road and to the part thereof so constructed as if such application had been made, and all the proceedings of such company had been had pursuant to the provisions of the said chapter. Provision in case public highways are used.

§ 4. Nothing in this act contained shall be deemed or construed to authorize the laying out or construction of any road in the cases specified in section nine of said chapter two hundred and ten of the Laws of 1847, nor to authorize the bridging or obstructing of any stream navigable by vessels or steamboats. Saving clause.

§ 5. This act shall take effect immediately.

Chap. 360.

AN ACT *to amend an act entitled "An act to provide for the incorporation of companies to construct plank roads, and companies to construct turnpike roads," passed May 7th,* 1847.

Passed April 12, 1848.

The People of the State of New York, represented in Senate and Assembly, do enact as follows:

Outer limits of the road not to exceed four rods.

§ 1. The commissioners appointed by the board of supervisors, as provided in the eighth section of the act to provide for the incorporation of companies to construct plank roads, and of companies to construct turnpike roads, passed May 7th, 1847, are hereby authorized in laying out a plank road, to determine the distance that the outer limits of the road shall be apart, as they may judge necessary, provided, in no case shall the company take more than four rods in width, except by the voluntary sale of the same to the company.

Half rates of toll.

§ 2. Any company formed under this said act, may take half the rates of tolls, and no more, provided for in said act, from persons living within one mile of the gate at which it is taken ; but no tolls shall be taken from farmers going to and from their work on their farms. [*See* § 4, *chap.* 546, *Laws of* 1855.]

Chap. 107.

AN ACT *in relation to plank roads and turnpike roads.*

Passed April 9, 1851.

The People of the State of New York, represented in Senate and Assembly, do enact as follows:

Who exempt from tolls.

§ 1. The following persons, and no others, shall be exempt from the payment of tolls at the gates of the several plank road companies formed under the act entitled "An act to provide for the incorporation of companies to construct plank roads, and for companies to construct turnpike roads," passed May seventh, one thousand eight hundred and forty-seven:

1. Persons going to or from religious meetings held at the place where such persons usually attend for religious worship, in the town where they reside or an adjoining town, or within eight miles of their residence:

2. Persons going to or from any funeral, and all funeral processions:

3. Troops in the actual service of this state, or of the United States, and persons going to or from a militia training, which by law they are required to attend:

4. Persons going to any town meeting or general election, at which they are entitled to vote, for the purpose of voting or returning therefrom:

5. Persons living within one mile of any gate by the most usually traveled road, shall be permitted to pass the same at one-half the usual rates of toll, when not engaged in the transportation of other persons or the property of other persons:

6. Farmers living on their farms, within one mile of any gate, by the most usually traveled road, shall be permitted to pass the same free of toll, when going to or from their work on said farms.

§ 2. It shall not, at any time hereafter, be lawful for any plank road company formed under the said act of May seventh, eighteen hundred and forty-seven, or for any turnpike company to erect or put up any toll gate, gate house, or other building, within a less distance than ten rods from the front of any dwelling house, barn, or other out house, without the written consent of the owner thereof; and if any toll gate or other such building shall hereafter be located by any such company within said distance, without such consent, the county judge of the county in which such building shall be so located, shall, on application, order the same to be removed. Gates, how located.

§ 3. Anything contained in the said act of May seventh, eighteen hundred and forty-seven, or in any subsequent act which is inconsistent with the provisions of this act, is hereby repealed.

Chap. 245.

AN ACT *in relation to plank roads.*

Passed April 15, 1853.

The People of the State of New York, represented in Senate and Assembly, do enact as follows:

Rates of toll. § 1. Instead of the toll authorized to be demanded and received on plank roads, by section thirty-five of the act entitled "An act to provide for the incorporation of companies to construct plank roads, and of companies to construct turnpike roads," passed May 7, 1847, the following rates of tolls may hereafter be demanded and received: for every vehicle drawn by one animal, one cent per mile, and one cent per mile for each additional animal; for every vehicle used chiefly for carrying passengers, drawn by two animals, three cents per mile, and one cent per mile for each additional animal; for every horse rode, led or driven, three-quarters of a cent per mile; for every score of sheep or swine, one and a half cents per mile; and for every score of neat cattle, two cents per mile.

§ 2. Sections twelve and thirteen of title four, chapter thirteen, part one, of the Revised Statutes, shall apply to plank road companies.

§ 3. This act shall take effect immediately.

Chap. 481.

AN ACT *to amend an act entitled "An act in relation to plank roads," passed April* 15, 1853.

Passed June 30, 1853.

The People of the State of New York, represented in Senate and Assembly, do enact as follows:

§ 1. The provision of the said act in the following words, to wit: "for every vehicle used chiefly for carrying passengers, drawn by two animals, three cents per mile, and one cent per mile for each additional animal," shall not apply to any plank road company located wholly or in part, in the counties of Chenango, Otsego, St. Lawrence and Oneida.

Certain provisions not to apply to roads in Chenango, Otsego, St. Lawrence and Oneida.

§ 2. This act shall take effect immediately.

Chap. 626.

AN ACT *to amend an act to provide for the incorporation of companies to construct plank roads, and of companies to construct turnpike roads, passed May* 7, 1847, *and the acts amendatory thereof.*

Passed July 21, 1853.

The People of the State of New York, represented in Senate and Assembly, do enact as follows:

§ 1. Every person liable to do highway labor, living or owning property on the line of any plank road of this state, may, on making application in writing to the commissioner or commissioners of their respective towns, on or any day previous to the time of making the highway warrants by such commissioners, be assessed the apportionment of highway labor, for such property upon such plank road, and the commissioner or commissioners shall

Highway labor, how assessed.

assess such person for the land or property owned by him, in or upon the line of said plank road as a separate road district. [*As amended by chap.* 495, *Laws of* 1855.]

Duty of highway commissioners.

§ 2. It shall be the duty of the highway commissioner or commissioners of such town to make a separate list of such persons and such land or property so assessed, as commissioners are now by law required to make for every separate road district, which shall be delivered to some one of the directors of such road, who shall proceed to have said highway labor worked on such road, in the same manner that overseers of highways are required by law to do.

Powers of directors.

§ 3. The said directors shall possess all the powers and have the same authority to compel the performance of such highway labor, or the payment of such highway tax as the overseers of highways now have by law, and shall make like return to the commissioners of highways.

May commute.

§ 4. Any person so assessed may commute for the tax, assessed upon him or his property, by paying the sum now fixed by law to any of said directors.

§ 5. This act shall take effect immediately.

Chap. 87.

AN ACT *in relation to plank roads and turnpike roads.*

Passed March 28, 1854.

The People of the State of New York, represented in Senate and Assembly, do enact as follows:

Road may be surrendered.

§ 1. The directors of any plank road company or turnpike company, formed under the act passed May 7, 1847, entitled "An act to provide for the incorporation of companies to construct plank roads, and for companies to construct turnpike roads," and every plank road company or turnpike company incorporated under or by any law of this state, are hereby authorized to abandon the whole or any part of their plank road or turnpike road, at either or both ends thereof, whenever the stockholders holding two-thirds of the stock in said road company shall consent to the same by a written declaration of the surrender of such part or parts of said road, which said declaration shall be attested by their common seal, and acknowledged by the president and secretary of said company before an officer empowered to take the acknowledgments of deeds. Such declaration and consent shall be filed and recorded in the clerk's office of the county in which the part or parts of said road abandoned shall be situated, and thereupon the plank or turnpike road, or the portion thereof so surrendered, shall cease to be the road or property of the company, and revert and belong to the several towns through which it was constructed, and the said company shall be no longer bound to maintain it or be liable to be assessed thereon, or be permitted to collect tolls for traveling over the same, from the time of recording said declaration of surrender and consent, without impairing the right of said company to take toll on the remaining part of their plank road or turnpike road at the rate prescribed in its charter, or by the laws of this state relating to any such company.

§ 2. Any plank road or turnpike company within this state, which shall have once laid their road with plank, may hereafter relay the same, or any part thereof, with broken stone, gravel, shells or other hard materials, whereby they keep a good and substantial road, but such plank road company shall collect only turnpike tolls on such part of their road as is constructed with broken stones, gravel, shells or other hard materials. [*See* § 7, *chap.* 546, *Laws of* 1855.] **Road may be relaid with gravel, &c.**

§ 3. Section thirty-five and thirty-six of the before mentioned act, passed May 7, 1847, is hereby amended by striking out, after the word "roads" in the third line of said section, "but not within three miles of each other." **Gates.**

Exemption from taxes.

§ 4. Toll houses and other fixtures and all property belonging to any plank or turnpike road company, shall be exempt from assessment or taxation for any purpose whatever; but no company shall be exempted whose net annual income, over and above all expenses of repairs and collection of tolls, is equal to 5 per cent on the original cost of the road. [*See* § 5, *chap.* 546, *Laws of* 1855.]

Bridges.

§ 5. The directors of any plank road company may put up and maintain at conspicuous places at each end of any bridge on their road, the length of whose span is not less than twenty-five feet, a notice with the following words in large characters: "One dollar fine for riding or driving over this bridge faster than a walk."

Whoever shall ride or drive faster than a walk over any bridge upon which such notice shall have been placed and shall then be, shall forfeit for every offense the sum of one dollar.

Want of legal organization not to work forfeiture.

§ 6. Every company formed or organized under the act entitled "An act for the incorporation of companies to construct plank roads, and of companies to construct turnpike roads," passed May 7, 1847, and the several acts amending the same, shall be deemed to be a valid corporation, although such company may not have complied with the requirement of such acts in the formation and organization of such company, and preparatory to the construction of its road and no act or omission on the part of any such company or of its stockholders or officers, shall work a forfeiture of its corporate powers or franchises, unless the same was willful and malicious; but this section shall not affect or impair any right of action heretofore accrued.

§ 8. This act shall take effect immediately.

NOTE.—Section 7 was repealed by chapter 104 of Laws of 1855.

Chap. 485.

AN ACT *in relation to turnpikes and plank roads, and to prevent encroachments thereon.*

Passed April 14, 1855.

The People of the State of New York, represented in Senate and Assembly, do enact as follows:

§ 1. Any person who shall draw or haul, or cause to be hauled or drawn, any logs, timber, or other material, upon the road bed of any plank road or turnpike road, unless the same be entirely elevated above the surface of the road on wheels or runners, by which said road bed shall be injured, or who shall do or cause to be done any act by which said road bed or any ditch, sluice, culvert or drain appertaining to any turnpike or plank road, shall be injured or obstructed, or shall divert or cause to be diverted any stream of water so as to injure or endanger any part of any such turnpike or plank road, shall forfeit and pay the sum of five dollars as a penalty, in addition to the damages resulting from such wrongful act.

§ 2. Any person who shall designedly place or leave, or caused to be placed or left, any log, timber, wood, stone or other material, upon the land held by any turnpike or plank road company, for highway purposes, in such a way as to obstruct the travel upon such road, or to endanger property or persons passing upon such road, shall, in case he or she do not remove such obstruction within forty-eight hours after receiving a written notice from one of the directors of the company owning the road upon which such obstructions have been placed or left, forfeit the sum of ten dollars for every twenty-four hours such obstruction shall remain after such notice.

§ 3. Any person who shall pass any turnpike or plank road gate without paying the toll required by law, and with intent to avoid the payment thereof, shall for each offense forfeit and pay to the corporation injured thereby, ten dollars. The penalties in this and the preceding sections may be sued for and recovered by any company injured thereby, in any court having jurisdiction thereof.

§ 4. Whenever the president or secretary of any turnpike or plank road company, shall notify any inspector of roads in the county where such roads are situated, that any person is erecting or has erected any fence or other structure, upon any part of the premises set apart by due course of law, for any turnpike or plank road, the said inspector shall proceed to examine the facts, and if it shall appear that such fence or other structure is upon any part of any such road, the said inspector shall order the same to be removed; and any person who shall neglect or refuse to remove the same within twenty days, or such further time not exceeding three months, as may be fixed by the said inspector, shall forfeit and pay the sum of five dollars for every day during which said fence or other structure shall remain on the said road, to be sued for and recovered by the corporation owning such turnpike or plank road, in any court having jurisdiction thereof; provided that the said inspector shall not order the removal of any fence previously erected, between the first day of December and the first day of April.

§ 5. All acts and parts of acts inconsistent with this act are hereby repealed.

§ 6. This act shall take effect immediately.

Chap. 546.

AN ACT *in relation to plank roads and turnpike roads.*

Passed April 18, 1855; three-fifths being present.

The People of the State of New York, represented in Senate and Assembly, do enact as follows:

§ 1. In any action hereafter brought by or against any plank or turnpike road company organized under the laws of this state, which shall have been in actual operation, and being in the possession of a road upon which they have taken toll for five consecutive years next preceding the commencement of such action, parol proof of such corporate existence and use shall be sufficient for all purposes of the action, unless the opposing party shall set up a claim in his complaint or answer, duly verified, of title in himself to the road, or some part thereof, stating the nature of his title and right to the immediate possession and use thereof.

§ 2. No plank road company shall be deemed to have forfeited any privilege or franchise by reason of not having completed their road the whole distance mentioned or described in their articles of association.

§ 3. Every instrument in writing, purporting to be an agreement between any plank road company and the supervisor or commissioners of highways of any town, in pursuance to section twenty-six of chapter two hundred and ten of the Session Laws of eighteen hundred and forty-seven, and heretofore filed or recorded in any town clerk's office, shall be deemed and taken in all courts and places to be as valid and effectual an agreement as if the same had been made and executed at a regular meeting of such supervisor and commissioner or commissioners of highways. The provisions of this section shall not affect suits now commenced.

§ 4. Section second of chapter three hundred and sixty of the Session Laws of eighteen hundred and forty-eight, is hereby amended so as to read as follows: Any company formed under the said act, may take half the rates of toll, and no more, provided for in said act, from persons living within one mile of the gate at which it is taken, except persons residing in a city or incorporated village, who shall pay full toll; but no toll shall be taken from farmers going to or from their work on their farms, on which they reside, or from persons driving or leading animals to or from the pasture or field where they are usually kept.

§ 5. Section four of chapter eighty-seven of the Session Laws of eighteen hundred and fifty-four, is hereby amended so as to read as follows:

Toll houses and other fixtures, and all property belonging to any plank or turnpike road company, shall be exempt from assessment and taxation for any purpose whatsoever, until the surplus annual receipts of tolls on their respective roads, over necessary repairs and a suitable reserve fund for repairs and relaying of plank, shall exceed seven per cent per annum on the first cost of such road. In case of any disagreement between the assessors of any town, village or city, and any such company, concerning such exemption claimed, said company may appeal to the county judge of the county in which such assessment is proposed to be made, who shall, after due notice to the appealing party of such appeal, examine the books and vouchers of such company, and take such further proof as he shall deem proper, and shall decide whether such company is liable to taxation under this section, and his decision shall be final.

§ 6. No company organized under the act entitled "An act for the incorporation of companies to construct plank roads, and of companies to construct turnpike roads," passed May seventh, eighteen hundred and forty-seven, and the several acts amending the same, shall be deemed to have forfeited any of its corporate powers or franchises by reason of the omission of the inspectors of elections for director in any such company to take the oath prescribed, prior to holding said election.

§ 7. Section two of chapter eighty-seven of the Laws of eighteen hundred and fifty-four, is hereby amended so as to read as follows:

Any plank road company or turnpike company within this state, which shall have once laid their road with plank, may hereafter relay the same or any part thereof with broken stone, gravel, shells, or other hard material, whereby they keep a good and substantial road. Such company shall be entitled to collect and receive the same tolls as is provided by chapter two hundred and forty-five of the Laws of eighteen hundred and fifty-three.

§ 8. The treasurer of every plank road company and turnpike company shall at the end of each fiscal year of said company, make and prepare under oath, a statement of the affairs of said company, in which he shall state the amount received by said company during the year, and from what sources the same was received, stating the amount received from each source separately; and also the amount expended during the year, and on what account the expenditures were made, and the items of said expenditures, and shall also state the amount of liabilities of said company, and amount of indebtedness to said company. Which statement he shall exhibit at all seasonable hours to any stockholder in said company, on being requested to do so; and in case such treasurer shall refuse to exhibit such account or statement as aforesaid, to any stockholder on request as aforesaid, he shall forfeit and pay to the person making such request the sum of five dollars for each offense, to be recovered in any court having cognizance thereof.

§ 9. This act shall take effect immediately.

Chap. 202.

AN ACT *in relation to plank roads and turnpike roads.*

Passed March 31, 1857.

The People of the State of New York, represented in Senate and Assembly, do enact as follows:

Stockholders may be directors.

§ 1. Whenever the whole number of stockholders in any plank road company or turnpike road company, shall not exceed the number of directors specified in the articles of association of such company, each stockholder shall be in fact and in law a director of such company, and in such case the stockholders shall constitute the board of directors, whatever may be their number, and a majority thereof shall form a quorum for the transaction of business.

§ 2. This act shall take effect immediately.

Chap. 482.

AN ACT *in relation to the sale of plank roads and turnpike roads on execution, and to provide for the incorporation of the purchasers at such sales into companies to own and operate such roads.*

Passed April 15, 1857.

The People of the State of New York, represented in Senate and Assembly, do enact as follows:

Operating roads.

§ 1. Whenever any plank road or turnpike road shall be sold upon any execution, and shall not be redeemed from such sale, according to law, then it shall and may be lawful for the purchaser or purchasers of such road, and they are hereby authorized to maintain and operate the same, in the same manner and subject to the same privileges and restrictions in all respects as the company owning such road at the time such sale was made.

§ 2. Such purchaser or purchasers, on associating with him or them not less than four persons, may be formed into a corporation for the purpose of owning such plank road or turnpike road, by complying with the following requirements: Articles of association.

They shall severally subscribe articles of association, in which shall be set forth the name of the company, the number of years the same is to continue, which shall not exceed the unexpired term of the original incorporation of the company whose road was so sold, whether it is a plank road or a turnpike road which the company is formed to own and operate; the amount of the capital stock of the company, which shall not exceed the amount of the capital stock of the company owning such road at the time of such sale; the number of shares of which the said stock shall consist; the number of directors, and their names, who shall manage the concerns of the company for the first year, and shall hold their offices until others are elected; the place from and to which said road is constructed, and each town, city and village into or through which said road shall pass, and its length as near as may be. Each subscriber to such articles of association shall subscribe thereto his name and place of residence, and the number of shares of stock owned by him in said company. The said articles of association may then be filed in the office of the secretary of state, and thereupon the persons who have so subscribed, and all persons who shall from time to time become stockholders in such company, shall be a body corporate, by the name specified in such articles, and shall possess the same powers and privileges and be subject to the same provisions as companies organized under the act entitled "An act to provide for the incorporation of companies to construct plank roads and of companies to construct turnpike roads," passed May seventh, one thousand eight hundred and forty-seven.

§ 3. The provisions of this act shall apply to all such sales, the right of redemption upon which either has heretofore expired or shall expire after the passage of this act.

§ 4. This act shall take effect immediately.

Chap. 209.

AN ACT *in relation to plank roads and turnpike roads.*

Passed April 9, 1859; three-fifths being present.

The People of the State of New York, represented in Senate and Assembly, do enact as follows:

May construct branches.

§ 1. The directors of any plank road company or turnpike road company, formed under the act passed May seventh, eighteen hundred and forty-seven, entitled "An act to provide for the incorporation of companies to construct plank roads, and for companies to construct turnpike roads," may, with the written consent of the persons owning two-thirds of the stock, construct branches to their main line of road, or extend or change the route of their road, or any part thereof, whereby the public interest will be promoted, through any uncultivated or unimproved lands.

May take and hold necessary real estate.

§ 2. The directors of any such company may purchase, take and hold any real estate necessary for the aforesaid purposes, and by their agents, servants or other persons employed, may enter upon the lands of any person or persons, which may be necessary for said purpose, and may construct their road upon any lands so entered upon, purchased or held.

Lands to be surveyed.

§ 3. Before entering, taking or using any land for the purpose of this act, the directors of any such company shall cause a survey and map to be made of the lands intended to be taken or entered upon, for any of said purposes, and by which the land of each owner and occupant intended to be taken and used shall be designated, and which map shall be signed by the surveyor or engineer making the same, and by the president of such company, and acknowledged by them, and be filed in the office of the clerk of the county. The directors of any such company, by any of its officers, agents or servants, may enter upon any lands for the purpose of making any examination, and of making survey and map, doing no unnecessary damage.

§ 4. In case the directors of any such company cannot agree with the said owners and occupants of any land intended to be taken and used for the purposes of this act, the directors may apply to the judge of the county court for the appointment of three disinterested persons, not the owners of real estate in any town through which any land intended to be used for the purposes of this act, or in any town adjoining such town, as commissioners by whom the compensation to be paid for the damages suffered or to be suffered by any person or persons, by reason of taking any of said lands for the purposes of this act, shall be ascertained and determined, and in case of the death, resignation, refusal or disability to act of any of said commissioners, the said judge may appoint others in their place. The commissioners shall give at least ten days' written notice of the time and place to hear the parties interested, to be served personally on the parties interested, or in their absence from their dwellings or places of business, by leaving the same thereat, with some person of suitable age, and in case of any legal disability of such owner or owners, to act thereupon, serving notice in like manner, upon his or her guardian, or person appointed to act for him or her, as hereinafter directed, and in case any of said owners shall be married women, insane, infants or idiots, the said judge shall appoint some suitable person to attend in their behalf, before the said

Disagreement.

Commissioner to give notice.

commissioners, and take care of their interest in the premises. The commissioners may issue subpœnas to compel the attendance of witnesses to testify before them, and they, or any of them, may administer the usual oath to such witnesses. They shall determine the width of road through said lands, and make a report of all proceedings before them, containing the testimony taken by them, and make an actual survey and description thereof, as laid out by them, and the sum awarded to each owner or any other person, duly signed or acknowledged by them, and return the same to said judge to be filed on record.

Compensation.

§ 5. Each commissioner is entitled to receive two dollars per day for his fees, to be paid by the company.

Appeal

§ 6. The directors of any plank or turnpike road company, or any party to the proceedings of the commissioners, may appeal from any award or determination of the commissioners, to the said county judge, providing the party appealing shall, within ten days after such award or determination shall be made, give written notice of the appeal to the other party or parties interested in the same; and the said judge shall examine the report of the commissioners, and if their proceedings in the case have been irregular, the said judge may set the same aside and order new proceedings and appointments; and the said judge may make such orders in reference to the proceedings of the commissioners, and of notices to be given by the parties as may not be inconsistent with this act, and as the nature of the case, and the interest of the parties may require. And the said commissioners shall again examine the case, and the decision then made shall be final.

Upon payment of compensation, may enter upon land.

§ 7. Upon the payment or legal tender of the compensation determined as before provided, the said directors of any plank road company, or any turnpike road company, shall be entitled to enter upon for the purposes contemplated by this act, all the lands and real estate for which such compensation shall be paid or tendered, as aforesaid, and to hold and use the same for the said purposes, to them and their successors forever. If any person to whom any compensation shall be awarded, or who shall be entitled to the same by virtue of said award, cannot be found, or shall refuse to receive the sum awarded to him or her, then the said payment may be made by depositing the amount of the said award to the credit of said person, in such bank as may be appointed by said judge. If the person to whom compensation is awarded, or who is entitled to receive the same as aforesaid, be under legal disability as aforesaid, payment may be made to his guardian or person appointed as aforesaid, by said judge, and if said guardian or person appointed cannot be found, then by deposit in bank as aforesaid.

Directors to take and hold real estate.

§ 8. The directors of any plank road company or turnpike road company, shall take and hold, for the purpose contemplated in this act, all the lands and real estate which they shall in any way legally enter upon and take by virtue hereof, to them and their successors so long as the same shall be used for a road.

Rights of purchaser of plank roads.

§ 9. Any person or persons who have heretofore or may hereafter purchase any plank road or any part thereof exceeding three miles under and by virtue of any mortgage executed by any plank road company incorporated under the general plank road law of this state, shall become the owner or owners of said road or part thereof thus purchased, and all the rights, privileges and franchises belonging to such plank road company at the time of such purchase, subject to the same restrictions as now exist by law.

§ 10. All acts and parts of acts, so far as they are inconsistent with this act, are hereby repealed.

§ 11. This act shall take effect immediately.

FORMS AND INSTRUCTIONS.

PREPARED IN PURSUANCE OF A CONCURRENT RESOLUTION OF THE LEGISLATURE OF 1860.

☞ The references are to the sections as arranged in this pamphlet, and the numbers follow the order of the provisions of the statutes, as they occur in this arrangement.

☞ The forms are prepared for three commissioners. Where there is but one commissioner, the form can be varied to conform to the fact.

No. 1.

Article 1, Section 1, Subdivision 3.

Order Amending Description of Road.

Whereas a road, used as a highway in the Town of *Portland in the County of Chautauqua, leading from the Erie road between the dwelling houses of Elijah Fay and John R. Coney, to the old Erie road on the east line of Elisha Fay's farm,* was laid out by the commissioners of highways of the said town, on the 10*th day of June*, 1825, but not sufficiently described:

Now, therefore, the undersigned, commissioners of highways of the said town [*if all the commissioners meet but do not all sign the order, say* all the commissioners being present, *or if all do not meet, say* all the commissioners having been notified to meet at this time and place for the purpose], having met *at the dwelling house of John Adams* in the said town, for the purpose of causing said road to be ascertained, described and entered of record in the town clerk's office, and having caused a survey of the said road to be made, do order that the said road be and the same is hereby ascertained and described according to the said survey, beginning at [*here insert the description according to the survey*].

TIMOTHY JUDSON,
DARIUS TALLMAN,
RUSSELL FITCH,
Commissioners of Highways.

Portland, July 1, 1860.

No. 2.

Article 1, Section 1, Subdivision 3.

Order ascertaining and describing road used 20 years, but not recorded.

Whereas a road in the Town of *Portland in the County of Chautauqua, leading from the town line road, near the dwelling house of Ahira Hall, by the north line of the farm of Lloyd Burr, to the old Erie road, east of the dwelling house of Horace Adams,* has been used as a highway for twenty years, but not recorded:

Now therefore, [the rest like form No. 1.]

No. 3.

Article 1, Section 1, Subdivision 5.

Order dividing town into road districts.

The undersigned, commissioners of highways of the Town of *Arkwright in the County of Chautauqua* [*here insert the attendance of, or notice to, all the commissioners, as in form No.* 1], do hereby order that the said town be, and the same is hereby, divided into *thirteen* road districts, in the following manner, viz.:

District No. 1, shall comprise all that part of the said town lying *north of the south line of the road leading from the dwelling house of Abiram Orton, to the east line of said town*, and all the inhabitants residing in said district, *and all those residing on the said road above mentioned, liable to work on highways, and Porter Phelps and Levi Baldwin, and all persons residing with them or on their farms, and liable to work on highways*, are assigned to the said district No. 1.

District No. 2, shall comprise, &c. [*till the thirteen districts are described and all the inhabitants properly assigned to the several districts*].

WILLIAM JOHNSON,
RUSSELL W. MATTOON,
LUTHER CLOUGH,
Commissioners of Highways.

Arkwright, February 15, 1860.

No. 4.

Article 1, Section 3.

Commissioners' Annual Account.

The commissioners of highways of the Town of Auburn, in the County of Cayuga, render to the board of town auditors of the said town the following account, viz.:

1. The highway labor assessed in said town for the year ending the 28*th day of February*, 1856, was *sixteen hundred and thirty-nine* days, and the highway labor performed in said town during the said year, was *fourteen hundred and sixty-seven* days.

2. The said commissioners have received during the said year, the following sums of money for fines and commutations, and under the provisions of the statute in relation to highways and bridges, viz.:

1856.			
March, 1,	From former commissioners of highways,...........	Bal. on hand by their last annual account, ...	$165 73
" 3,	John Ray,...............	Bal. on hand as overseer, ..	3 84
June 5,	Abel Smith, ...	Commutation for $6\frac{1}{2}$ days' work,...............	4 06

3. During the said year the road *for one mile and sixty rods south of the Village of Auburn* has been turnpiked and graveled, the bridge *over the Owasco outlet, north of the Village of Auburn,* has been rebuilt [*describe the other improvements*], and all the roads in said town are now in good condition, except needing the ordinary spring repairs, and all the bridges safe and apparently permanent [*make a true description of the state of roads and bridges in a brief manner*].

4. The following improvements are necessary to be made on the roads and bridges in the said town during the present year, viz.: [*state the improvements plainly and briefly*] and the probable expense of making such improvements beyond what the labor to be assessed during the year will accomplish, is estimated at $

HENRY BRUCE,
JOEL KNOX,
WILLIAM TAYLOR,
Commissioners of Highways.

Auburn, Feb. 28, 1856.

No. 5.

Article 1, Section 4.

Statement and estimate of improvements for Supervisor.

To the supervisors of the Town of *Auburn, in the County of Cayuga:*

The commissioners of highways of the said town, render the following statement of the improvements necessary to be made in the roads and bridges in the said town, together with the probable expense thereof, viz.:

The *middle bridge across Owasco creek* needs *a new butment and new plank,* the probable expense of which will be $85.00, to wit: *for the butment* $40.00, *and for the plank,* 2,250 *feet, at* $20.00 *per M.*, $45.00. [*State the other items with equal particularity.*]

HENRY BRUCE,
JOEL KNOX,
WILLIAM TAYLOR,
Commissioners of Highways.

Auburn, Oct. 20, 1856.

No. 6.

Article 1, Section 8.

Additional assessment by Overseer.

Whereas the quantity of labor assessed on the inhabitants of road district No. 3, *in the Town of Schoharie in the County of Schoharie,* by the commissioners, is deemed insufficient by me, *Philip Lasher,* overseer of highways of said district, to keep the roads therein in repair. I do hereby make another assessment according to the form of the statute in such case made and provided, on the actual residents in such district, in the same proportion, as near as may be, and not exceeding one-third of the number of days assessed in the present year by the commissioners on the inhabitants of such district which additional assessment is as follows, viz.:

NAMES OF INHABITANTS.	ASSESSMENT BY COMMISSIONERS.	ADDITIONAL ASSESSMENT.
William Winters, ...	Days, 42	Days, 7
Peter Shafer, ...	" 9	" 2

PHILIP LASHER,

Overseer of District No. 3.

Schoharie, Feb. 7, 1860.

No. 7.

Article 1, Section 11.

Direction to Overseer to procure Scraper or Plow.

The undersigned, commissioners of highways of the Town of *Schoharie in the County of Schoharie,* deeming it necessary and useful, hereby direct and enforce *Philip Lasher,* overseer of highways of road district No. 3 in said town to procure a good and sufficient iron [*or steel*] shod scraper, and a plow [*or either separately*], for the use of his road district, to be paid for by the moneys arising from commutations and fines within such district.

ELI GARDINER,
TOBIAS BOUCK,
ORSON ROOT,
Commissioners of Highways.

Schoharie, Oct. 21, 1860.

No. 8.

Article 1, Section 12.

Assessment by Overseer for Scraper or Plow.

Whereas the commissioners of highways of the Town of *Schoharie in the County of Schoharie*, on the *1st day of November*, 1859, by writing under their hands, directed and empowered me, *Philip Lasher*, overseer of highways of road district No. 3 in said town, to procure a good and sufficient iron [*or steel*] shod scraper and a plow [*or either separately*], for the use of my said road district, to be paid for by the moneys arising from commutations and fines within such district; and whereas such moneys are insufficient for the purpose, by the amount of $8.50:

Now, therefore, I, the said overseer, according to the form of the statute in such case made and provided, do hereby assess the deficiency of eight dollars and fifty cents aforesaid, upon the inhabitants of the said district, in the proportion they are respectively assessed on the assessment roll of said town; which said assessment is as follows, viz.:

NAMES OF INHABITANTS.	TOWN ASSESSMENT.	OVERSEER'S ASSESSMENT.
Old Abe Lincoln, ...	$30 94	$2 87
Steve Myers, ...	17 90	1 79

PHILIP LASHER,
Overseer of District No. 3.

Schoharie, Nov. 1st, 1860.

No. 9.

Article 1, Section 14.

Warrant appointing Overseer in case of vacancy.

ALBANY COUNTY, } ss.
Town of Knox, }

Whereas a vacancy exists in the office of overseer of highways in road district No.   in the said town, by reason of the [refusal to serve, death, removal, &c., &c., according to the fact] of Luther Crocker, who was elected to the said office:

Now, therefore, pursuant to the statute in such case made and provided, we the subscribers, commissioners of highways of the said town, [in case all the commissioners do not sign the warrant, say all the commissioners having been notified to attend at this time and place, for the purpose of filling said vacancy,] having met for the purpose of filling such vacancy, do hereby appoint Ira Porter, overseer of highways for road district No.   in the said town of Knox.

A. B.
C. D.
E. F.

Knox, April 15, 1856.

No. 10.

Article 1, Section 17.

Complaint to Commissioners against an Overseer.

To the commissioners of highways of the Town of *Schoharie*, in the County of *Schoharie*:

The complaint of *Frederick Rowley*, a resident of the said town, respectfully showeth,

That *Peter Lasher*, overseer of highways of road district No. 3 in the said town, has refused and neglected to warn *B. V. Kniskern* to work on the highways according to the assessment upon him, although he has been required so to do by the commissioners [*or one of the commissioners*].

FREDERICK ROWLEY.

Schoharie, June 20, 1859.

☞ If the neglect complained of be different from the above, state it distinctly in the complaint. A great variety of cases may occur.

No. 11.

Article 1, Section 17.

Security for prosecuting Overseer.

Whereas *Frederick Rowley* has made complaint to the commissioners of highways of the Town of *Schoharie*, that *Peter Lasher* overseer of highways of road district No. 3 in said town, has refused and neglected [*here insert the matter complained of, as in the complaint*]. Now, therefore, we, *Frederick Rowley* and *Eli Gardner*, do hereby covenant, promise and agree, with the said commissioners of highways, that we will well and truly indemnify and save them harmless against the costs which may be incurred in prosecuting for the penalty annexed to such refusal or neglect. Sealed with our seals, and dated the 20th day of June 1859.

FREDERICK ROWLEY. [L. S.]
ELI GARDNER. [L. S.]

No. 12.

Article 2, Section 21.

Overseer's List of Inhabitants liable to work on Highways.

I, A. B., overseer of highways of road district No. , in the Town of , in the County of , do hereby certify that the following is a list of all the inhabitants of my road district who are liable to work on the highways:

A. B. C. D. E. F.

A. B., *Overseer.*

, *March* 15, 1859.

No. 13.

Article 2, Section 22.

Commissioners' List and Statement of Non-resident Lands.

A list and statement of the contents of all lots, pieces or parcels of land, within the Town of , in the County of , owned by non-residents therein:

OWNERS' NAMES.	DESCRIPTION OF LAND.	VALUE ACCORDING TO LAST ASSESSMENT-ROLL.	NO. OF DAYS.	ROAD DISTRICT.
Jere'h Newcomb,	E. Pt, L. 14, T. 5, Range 12.	$1,250 00	7	5

A. B.,
C. D.,
E. F.,
Commissioners of Highways.

, *April* 1, 1859.

No. 14.

Article 2, Section 24.

Commissioners' Estimate and Assessment of Highway Labor.

The undersigned commissioners of highways of the Town of , in the County of [*here insert the attendance of, or notice to, all the commissioners, as in No.* 1], having met at , in said town, to ascertain, estimate and assess the highway labor to be performed in the said town the ensuing year, have ascertained, estimated, and assessed the same as follows, that is to say:

1. The whole number of days' work assessed for the year is *twelve hundred*, being at least three times the number of taxable inhabitants of the said town.

2. Every male inhabitant above the age of twenty-one years, (excepting ministers of the gospel and priests of every denomina-

tion, paupers, idiots, and lunatics) being *four hundred and fifty-three*, is assessed one day [*or two days, or one day and a half, &c.*].

3. The residue of such work, being *seven hundred and forty-seven* days, is assessed as follows, viz.:

NAMES.	NO. OF DAYS.	NAMES.	NO. OF DAYS.
A. B.	6	C. D.	$5\frac{1}{2}$

The lands in said town owned by non-residents, are assessed as follows, viz.:

NAMES.	DESCRIPTION.	VALUE.	NO. OF DAYS.
E. F.	N. Pt. L. 4, T. 3, Range 7.	$1,500 00.	6

G. H.,
I. K.,
L. M.,
Commissioners of Highways.

, *April* 1, 1859.

☞ The commissioners are to affix to the name of each person named in the list furnished by the overseers the number of days' work assessed upon such persons respectively, and add to the several lists the description and assessment of each tract of non-resident land in the respective road districts.

No. 15.

Article 2, Section 24, Subdivision 5, and Section 34.

Certificate of Commissioners to Overseer's Lists.

We certify, that the number of days affixed to each name and to each tract of non-resident land in the annexed list, is the true assessment made by us upon such persons and tracts respectively for the ensuing year.

A. B.,
C. D.,
E. F.,
Commissioners of Highways.

, *April* 1, 1859.

No. 16.

Article 2, Section 35.

Overseer's Assessment of Persons left out of List, and of New Inhabitants.

The following named persons having been left out of the list of persons assessed to work on the highways in road district No. , in the Town of , in the County of , [*or having become inhabitants of road district No. , in the Town of , in the County of , since the list of assessments of highway labor for said district were made.*] Now I, overseer of highways of said district, according to the form of the statute in such case made and provided, do hereby assess and rate the said persons in proportion to their real and personal estate, to work on the highways, as others rated by the commissioners on such list, subject to an appeal to the commissioners, which said assessment is as follows, viz.:

A. B., four days.
C. D., three days.

E. F., *Overseer.*

, *June* 10, 1859.

No. 17.

Article 2, Section 35.

Appeal from Assessment of Overseer.

To the commissioners of highways of the Town of , in the County of :

The subscriber, being a new inhabitant of road district No. , in said town [*or having been left out of the list for road district No. , in said town, by the highway commissioners*], and having been assessed by the overseer of said district days' labor on the highway, and conceiving himself aggrieved thereby, hereby appeals to the commissioners from said assessment.

A. B.

, *June* 10, 1859.

No. 18.

Article 3, Section 42.

Overseer's Notice to Agent of Non-Resident.

To *James Lusk*, agent of *Reuben Bell*, owner of lands in the Town of *Dresden*, in the County of *Washington:*

Take notice, that the said *Reuben Bell* is assessed *five* days' labor on the highway in road district No. 7 in said town, and that the said labor is required to be performed from the 5th to the 12th days of July next, on the road leading from [*describe the locality with reasonable accuracy*], in the said district.

JABEZ HYDE,
Overseer of District No. 7.

Dresden, June 29, 1859.

No. 19.

Article 3, Section 43.

Overseer's Notice in relation to Labor assessed on Lands of Non-Residents.

Notice is hereby given, that the highway labor assessed on the following described parcels of land in the Town of *Pike, Allegany* county, owned by non-residents, is required to be performed on the highway in road district No.   , in said town, [between   and   , or leading from   to   , describing the locality with reasonable accuracy,] at the time specified opposite the description of said parcels of land, respectively :

OWNERS' NAMES.	DESCRIPTION OF LANDS.	AMOUNT OF ASSESSMENT.	TIMES OF PERFORMING LABOR.
Jared Rust, ...	N. E. pt. L. 10, Range 5, T. 7, 120 acres.	$4\frac{1}{2}$ days.	June 5 to 10.

GEORGE COLE,
Overseer of Highways.

Pike, May 10, 1859.

No. 20.

Article 3, Section 50.

Overseer's Complaint for Idleness, Neglect, &c.

SENECA COUNTY, *ss.:*

A. B., overseer of highways of road district No. , in the Town of *Ovid*, in the said county, being sworn, says that on the 17th day of June, 1859, he gave C. D., who resides in said district, and is assessed to work on the highways therein, notice to appear on the 19th day of June aforesaid, with a [*state what kind of team or implements were required*], on the road [*state where*], to do such work; and that the said C. D. did not appear nor furnish any one in his stead [*or did not bring such team or implement as was required, stating it: or, when he so appeared, was idle, or hindered others from working—or whatever the complaint is*], and has not paid the commutation money for said work, nor rendered a satisfactory excuse for such neglect.

A. B.

Subscribed and sworn to before me,
June 21, 1859,

G. H., *J. P.*

No. 21.

Article 3, Section 51.

Summons.

SENECA COUNTY, *ss:*

To any constable of the Town of *Ovid*, in said county:

Whereas A. B., overseer of highways of road district No. , in the said town, has made complaint on oath before me, a justice of the peace of the said town, that C. D., a resident of said road district, and assessed to work on the highways therein, after being duly notified to appear on the 19th day of June, instant, [*state what team or implements were required*], to do such work; and that the said C. D. [*stating the matter of the complaint*]; and has not paid the commutation money nor rendered a satisfactory excuse.

You are therefore hereby required to summon the said C. D. to appear forthwith before me, at my office, in the said town, to show cause why he should not be fined according to law for such refusal [*or neglect, or misconduct*].

G. H., *J. P.*

Ovid, June 21, 1859.

No. 22.

Article 3, Section 52.

Judgment.

In the matter of the complaint of
A. B., overseer of highways,
vs.
C. D.

The said C. D. having been duly summoned to appear before me, G. H., the justice of the peace to whom the said complaint was made, to show cause why he should not be fined according to law for the refusal [*or neglect, or misconduct*], set forth in the said complaint, as appears by the return of N. U., a constable of the said town; and no sufficient cause having been shown by him, and twenty-four hours having elapsed since the service of the said summons, as appears by the said return of the said constable, I do therefore impose upon the said C. D, a fine of , according to the form of the statute in such case made and provided, together with for the costs of the proceedings under the said complaint.

G. H., *J. P.*

Ovid, June 23, 1859.

No. 23.

Article 3, Section 52.

Warrant to levy Fine.

SENECA COUNTY, *ss:*

To any constable of the Town of *Ovid,* in the said county:

You are hereby commanded to levy of the goods and chattels of C. D. *four dollars and eighteen cents;* being *one dollar* for a fine imposed by me for [*specify the neglect or misconduct*], as set forth in the complaint of A. B., overseer of highways of road district No.   , in the said town; and also *three dollars and eighteen cents* for the costs of the proceedings on said complaint: and bring the said sum of money before me without delay.

G. H., *J. P.*

Ovid, June 23, 1859.

No. 24.

Article 3, Section 56.

Overseer's List of Non-Resident Lands, on which Labor has not been performed.

The following is a list of all the lands of non-residents and of persons unknown, in the Town of *Bern, Albany* county, which were taxed on my lists, on which the labor assessed by the commissioners of highways for the year 1859 has not been paid, and the amount of labor unpaid.

OWNERS' NAMES.	DESCRIPTION OF LANDS.	ASSESSED VALUE.	LABOR UNPAID.
John Reed, ...	S. W. Pt. L. 7, Range 14, T. 9. 75 acres.	$225.00	$1\frac{1}{2}$ days.

JOSEPH MILLS,
Overseer of Highways of Road District No. 9 *in said Town.*

Bern, September 30, 1859.

Affidavit on such List.

ALBANY COUNTY, *ss:*

Joseph Mills, overseer of highways in road district No. 9, in the Town of *Bern,* in said county, being sworn, says, that he has given the notice required in the [thirty-third or thirty-fourth, or both, as the fact may be], section of Title I, Chap. 16, of the First Part of the Revised Statutes, and that the labor for which such land is returned has not been performed.

JOSEPH MILLS.

Subscribed and sworn to before me, September 30, 1859.

JACOB STOW, *J. P.*

No. 25.

Article 3, Section 60.

Overseer's Annual Account.

To the commissioners of highways of the Town of *Salem, Washington county:*

The undersigned, overseer of highways of road district No. 11, in the said town, pursuant to law, renders the following account:

1. The names of all persons assessed to work on the highways in the district of which he is overseer.

NAMES.	NO. OF DAYS.	NAMES.	NO. OF DAYS.
John Williams, ...	3	Job Frost, ...	5

2. The names of all those who have actually worked on the highways, with the number of days they have so worked.

NAMES.	NO. OF DAYS.	NAMES.	NO. OF DAYS.
Job Frost, ...	5	John Williams, ...	$1\frac{1}{2}$

3. The names of all those who have been fined, and the sums in which they have been fined.

NAMES.	SUMS.	NAMES.	SUMS.
John Williams, ...	$1 87$\frac{1}{2}$	Peter Jackson, ...	$0 75

4. The names of all those who have commuted, and the manner in which the moneys arising from fines and commutations have been expended by him.

NAMES.	DAYS.	AMOUNT.	NAMES.	DAYS.	AMOUNT.
James Johnson, ...	3½	$2 19	Ralph Clark, ...	1	$0 62½

Whole amount received for fines and commutations, as above stated, $5.44; which has been expended in [turnpiking the road from   to   , or in building a bridge across *Muskrat* brook, &c., as the fact is].

5. A list of all lands which he has returned to the supervisor for non-payment of taxes, and the amount of tax on each tract of land so returned [here copy the list delivered to the supervisor].

JOHN CRARY,
Overseer of Highways,
District No. 11.

Salem, Feb. 27, 1859.

Affidavit.

WASHINGTON COUNTY, ss.

John Crary, overseer of highways, of road district No. 11, in the Town of *Salem*, in said county, being sworn says, that the annexed account is true.

J. CRARY.

Subscribed and sworn to before me, February 27, 1859,
STEPHEN ROSS, *Com. of Highways.*

No. 26.

Article 4, Sections 65, 66.

Application to lay out a Road.

To the commissioners of highways of the Town of *Clay*, in the County of *Onondaga:*

The subscribers, residents of said town [or, owning lands in said town], and liable to be assessed for highway labor therein, hereby apply to the said commissioners of highways to lay out a new road, commencing at [here describe the proposed road] which proposed road will not pass through any enclosed, improved, or cultivated land, except the lands of L. M. and N. O., who consent thereto.

G. H.
R. I.
L. M.
N. O.

This form may be varied, to suit the occasion, for the alteration or discontinuance of a road.

No. 27.

Article 4, Sections 67-76.

Order laying out a Road.

The undersigned, commissioners of highways of the Town of *Pomfret*, in the County of *Chautauqua* [*if all the commissioners meet but do not all sign the order*, *say*, all the commissioners being present *or if all do not meet*, *say*, all the commissioners having been notified to meet at this time and place for the purpose], having met at the inn of *Seril Manton*, in the said town, to decide upon the application of *Charles J. Orton*, a resident of the said town, liable to be assessed for highway labor therein, for the laying out of the road hereinafter described (twelve reputable freeholders of the said town, convened and sworn, after public notice of six days at three of the most public places of the town, according to law, having certified that such highway is necessary and proper, and the said commissioners having caused notice in writing to be given to *Luther Crocker* and *Richard Arnold*, occupants of the land through which the road is to run, at least three days before the time of meeting, of the time and place at which they would meet to decide upon the said application), do order that a public highway, three rods wide, shall be and the same is hereby laid out, pursuant to the said application, the centre whereof is the following described line, viz.: Beginning [*here insert the survey*].

GEORGE C. ROOD,
DOW C. SMITH,
JOSIAH W. BRISTOL,
Commissioners of Highways.

Pomfret, July 1, 1859.

No. 28.

Article 4, Section 67.

Order to Alter a Road.

The undersigned, commissioners of highways of the Town of *Pomfret*, in the County of *Chautauqua* [*here insert the attendance of, or notice to, all the commissioners, as in the preceding*], having met at the dwelling house of *John Bartlett*, in the said town, to decide upon the application of *Anson Reed*, a resident of the said town, liable to be assessed for highway labor therein, for the alteration of the road between the chair-shop of *Norman Neff* and *Shumla* [*if the altered line of the road passes through improved lands and the owner's consent is not obtained, insert the same preliminary proceedings as in the preceding*], do order that the line of the said road be, and the same is hereby so altered as to run from a point feet south of the southeasterly corner of said *Neff's chair-shop*, in a straight line. South, degrees East, till it strikes the present centre of the road, thence along the present line thirty-one rods, and thence continuing in a straight line sixty-three rods to the centre of the present road, on the summit of the hill North of *Shumla*, the said line to be the centre of the road, which shall remain of the width of three rods. And it is further ordered, that such parts of the present road as are not included in the above description be and the same are hereby discontinued.

GEORGE C. ROOD,
DOW C. SMITH,
JOSIAH W. BRISTOL,
Commissioners of Highways.

Pomfret, July 1, 1859.

No. 29.

Article 4, Section 65.

Order to Discontinue a Road.

The undersigned, commissioners of highways of the Town of *Pomfret*, in the County of *Chautauqua* [*here insert the attendance of, or notice to, all the commissioners as in the preceding*], having met at the dwelling house of *Benjamin Griswold*, in the said town, to decide upon the application of *John Arnold*, a resident of the said town, liable to be assessed for highway labor therein, for the discontinuance of the road hereinafter described (twelve disinterested freeholders of said town, duly summoned and sworn, having certified in writing that the said road is useless and unnecessary), do order that the road, beginning at [*here insert a description or survey of the road*], in the said town, be and the same is hereby discontinued.

GEORGE C. ROOD,
DOW C. SMITH,
JOSIAH W. BRISTOL,
Commissioners of Highways.

Pomfret, July 1, 1859.

No. 30.

Article 4, Section 71.

Notice of Application to lay out Highway.

Notice is hereby given that the subscribers, persons liable to be assessed for highway labor in the Town of , in the County of , have applied to the commissioners of highways of the said town, to lay out a highway in said town, beginning [*insert a description of the proposed road*], which said highway is proposed to be laid through the improved lands of and [*if necessary for identity, describe the particular lots*], and that twelve reputable freeholders of the town will meet on the 20*th day of June instant, at* 10 *o'clock A. M., at the dwelling house of* , to examine the ground.

A. B.,
C. D.

, *June* 14, 1859.

No. 31.

Article 4, Section 73.

Freeholders' Certificate to lay out a Road.

The subscribers, freeholders of the Town of *Hampton, Washington* county, not interested in the lands through which the road hereinafter described is to be laid, nor of kin to the owners thereof, having met at the dwelling house of *Beman Austin*, in the said town, and having been first duly sworn, have personally examined the route of the highway proposed to be laid out by the application of *Joab Stow*, beginning at [*here insert a description or survey of the road*], and have heard and considered all reasons offered for or against laying out the same, and thereupon do hereby certify to the commissioners of highways of said town, that we are of opinion that such highway is necessary and proper.

MASON HULETT.
&c., &c.

Hampton, July 1, 1859.

No. 32.

Article 4, Section 73.

Commissioners' Notice to Occupant.

To *Mr. Richard Arnold:*

Sir—The commissioners of highways of the Town of *Pomfret* will meet at *Seril Manton's tavern*, in said town, on *Tuesday, the* 10*th day of July, instant, at* 10 *o'clock in the forenoon*, to decide on the application of *Charles J. Orton* for a highway to be laid out beginning [*here insert the description as in the application*], twelve reputable freeholders, summoned and sworn pursuant to statute, having certified that such highway is necessary and proper.

JOSIAH SMITH,
JOHN TARBOX,
JAMES HILTON,
Commissioners of Highways.

Pomfret, July 7, 1859.

No. 33.

Article 4, Section 75.

Release of Claim to Damages.

Whereas, application has been made to the commissioners of highways of the Town of , in the County of , to lay out a highway in said town beginning [*describe it as in the application*], which proposed highway will pass through my improved lands: Now, therefore, in consideration of the laying out of the said highway, I do hereby release all claim to damages by reason thereof.

Sealed with my seal, and dated the day of , 1859.

B. Y. [L. S.]

No. 34.

Article 4, Section 75.

Agreement of Owner and Commissioners, as to Damages.

Whereas, the commissioners of highways of the Town of , in the County of , have, by an order dated the day of , 1859, laid out a highway in said town, beginning [*describe it as in the order*], which said highway passes through the improved lands of J. H.: Now, therefore, the damages of the said J. H., by reason of the laying out of said highway, are hereby ascertained by agreement of the said J. H. and the said commissioners of highways, at the sum of *one hundred dollars.*

J. H.

B. L.,
Z. X.,
P. M.,
Commissioners of Highways.

, *July* 10, 1859.

No. 35.

Article 4, Section 76.

Application to County Court to appoint Commissioners to assess Damages.

To the Hon. the County Court of county:

Whereas, the commissioners of highways of the Town of , in said county, by an order dated , 1859, have laid out a highway in said town, beginning [*insert description as in the order*]:

Now, therefore, the undersigned, commissioners of highways of the said town, hereby apply for the appointment of commissioners to assess the damages occasioned by the laying out of said highway, pursuant to the statute in such case made and provided.

E. G.,
H. B.,
N. B.,
Commissioners of Highways.

, *June* 13, 1859.

No. 36.

Article 4, Section 80.

Assessment by Commissioners.

Whereas, the undersigned, , and , were appointed by an order of the County Court of the County of , made on the day of , 1859, on the application of E. G., H. B. and N. R., commissioners of highways of the Town of , in said county, commissioners to assess the damages occasioned by the laying out of a highway in the said town, beginning [*here insert description as in the order*], which highway passes through the improved lands of , and , and was laid out by the commissioners of highways of the said town, by an order, dated , 1859:

Now, therefore, we, the said commissioners, having taken the oath of office prescribed by the Constitution, and having all met and acted on the matter committed to us, at the in said town, this day of , 1859, pursuant to a notice from said commissioners of highways, of at least six days, according to law, and having taken a view of the premises, and heard the parties and such witnesses as have been offered before us; do thereupon, determine and assess the damages required to be assessed on the said highway as follows, viz.: We assess the damages of , at dollars. We assess the damages of , at dollars.

—— ——,
—— ——,
—— ——,
Commissioners.

, *July* 30, 1859.

No. 37.

Notice when Commissioners or Parties are Aggrieved or Dissatisfied.

Notice is hereby given, that I , conceiving myself aggrieved [*or we, the commissioners of highways, feeling dissatisfied*], by the assessment of damages made by , and commissioners appointed by the County Court of the County of , to assess the damages occasioned by the laying out of a highway in the Town of , in said county, beginning [*insert description*] which said assessment was filed in the office of the town clerk of the said town, on the day of , 1859, do hereby ask for a jury to re-assess the said damages, and such jury will be drawn at the clerk's office, of the Town of , in said county, adjoining the said Town of , on the day of , 1859 [*not less than* 10 *nor more than* 20 *days from the filing of the assessment*], by the town clerk of said Town of [*where drawn*].

, *June* 15, 1859.

To be signed by the person aggrieved, or the commissioners dissatisfied.

[This notice must be served on the town clerk and opposite party, within 20 days from the filing of the assessment, and must be served on the clerk first, and the other party within three days after; and all in season to have the drawing of the jury within the 20 days.]

No. 39.

Summons for Jury.

COUNTY, *ss.*

To , one of the constables of the Town of in the said county:

You are hereby directed to summon [*name the twelve persons*] to appear at , in the said town, on the day of , 1859, to make a jury to re-assess the damage occasioned by the laying out of a highway, in the said town, by the highway commissioners thereof, on the day of , 1859. Hereof fail not.

J. B., *J. P.*

No. 40.

Oath of Jury.

You and each of you do solemnly swear, in the presence of Almighty God, well and truly to determine and re-assess such damages as shall be submitted to your consideration.

No. 41.

Verdict of Re-Assessment.

We the subscribers, a jury duly drawn and sworn, to determine and re-assess the damages occasioned by the laying out of a highway in the Town of , in the County of , by the highway commissioners thereof, on the day of , 1859, which said highway runs from [*describe the highway as in the order, and state whose lands it passes through*], having taken a view of the premises and heard the parties and such witnesses as have been offered by them and sworn before us, do hereby determine and re-assess the said damages as follows, viz.: We determine and re-assess the damages of H. B. at dollars [*specify each person's damages passed upon.*]

To be signed by the six jurors.

, *July*, 1859.

COUNTY, *ss.*

I, J. B., the justice of the peace by whom the within [*or above*] named jury were summoned, drawn and sworn, do certify that the within [*or above*] is the verdict of re-assessment rendered by the said jury, pursuant to the said proceedings, this day of July, 1859.

J. B., *J. P.*

No. 42.

Article 4, Section 97.

Freeholders' Certificate to Discontinue a Road.

The subscribers, disinterested freeholders of the town of , in the County of , having met at the dwelling house of , in the said town, in pursuance of a summons from the commissioners of highways of the said town, to examine and certify in regard to the propriety of discontinuing the highway from [*here describe the highway to be discontinued*], and having been duly sworn, have viewed the said road, and thereupon do certify that we are of opinion that the same is useless and unnecessary.

——————,
——————,
&c., &c., &c.

, *July* 1, 1859.

No. 43.

Article 4, Section 100.

Appeal to the County Judge.

To the , County Judge of county:

I, , conceiving myself aggrieved by the determi nation of the commissioners of highways of the Town of , in the said County, made on the day of , in [*laying out, altering, discontinuing, refusing to lay out, alter or discontinue*], a highway in the said town, from [*describe the road as in the order of the commissioners*], upon the application of , do hereby appeal from the said determination of the said commissioners, and pray the appointment of referees according to the form of the statute in such case made and provided, to hear and determine my said appeal.

The ground upon which my appeal is made, is that [*here set forth the couse of complaint*] and my appeal is brought to reverse entirely the determination of the commissioners [*or if part only, then*]. And my appeal is brought only to reverse that part [*or those parts*] of the determination of the commissioners which [*here set forth the particular part or parts*].

—— ——.

No. 44.

Appointment of Referees.

COUNTY, *ss:*

Whereas, , of the Town of , in the said county, has appealed from the determination of the commissioners of highways of the said town, made on the day of , 18 , in [*laying out, altering, discontinuing, refusing to lay out, alter, discontinue*], a highway in the said town, which highway is particularly described in the said appeal hereto annexed [*and whereas, Richard Bailey and John Smith have also appealed from the same determination of the commissioners, which said appeals are also hereto an-*

nexed], and sixty days having elapsed after such determination has been filed in the office of the town clerk of the said town:

Now, therefore, I, , county judge of the said county, to whom the said appeal was [*or appeals were*] addressed, according to the form of the statute in such case made and provided, do hereby appoint James Johnson, of the Town of , , of the Town of , and , of the Town of *Geneseo*, three disinterested freeholders who have not been named by the parties interested in the appeal, and who are residents of the county, but not of the town wherein the road is located, as referees, to hear and determine all the appeals that have been brought in relation to the said determination of the said commissioners.

—— ——,
County Judge.

No. 45.

Order of Referees deciding an Appeal.

Whereas, of the Town of in the County of , on the day of , 18 , appealed to the county judge of the said county, from the determination of the commissioners of highways of the said town, made on the day of 18 , in [*laying out, altering, discontinuing, refusing to lay out, alter, discontinue.*] a highway in the said town, which highway is particularly described in the said appeal hereto annexed [*and whereas, Richard Bailey and John Smith, also appealed from the same determination of the said commissioners, which said appeals are also hereto annexed*], and whereas, after the expiration of sixty days after such determination had been filed in the office of the town clerk of the said town, the said county judge, according to the form of the statute in such case made and provided, appointed us, of the Town of of the Town of , and , of the Town of , three disinterested freeholders, who had not been named by the parties interested in the appeal, and who are residents of the county, but not of the town wherein the road is located, as referees, to hear and determine all the appeals that had been brought in relation to the said determination of the said commissioners, which said appointment is hereto annexed, and we having given notice pursuant to law, to the said commissioners of highways [*and to John Rogers an applicant for such road*], specifying the day of , as the time, and the dwelling house of , in said Town of , as the place, at which we would convene to hear the appeal, which notice was duly served at least eight days before the said time of convening, to wit, on the day of , 18 , and we having convened at the time and place specified in said notice, and, before proceeding to hear the said appeal [*or appeals*], having been duly sworn by an officer authorized to take affidavits to be read in courts of record, to wit, , a justice of the peace of the said county, faithfully to hear and determine the matters referred to us, have heard the proofs and allegations of the parties, and do thereupon order, determine, and adjudge that the said determination of the said commissioners of highways be, and the same is hereby reversed [or affirmed].

—— ——,
—— ——,
—— ——,
Referees.

No. 46.

Order of Referees when a road is laid out by them.

The same form as No. 45 to the word "*reversed.*"

And we do further order, determine, and adjudge, that in pursuance of the said application of the said John Rogers, a highway be and the same is hereby laid out, beginning [*here insert description of the road, and if the description is by giving the centre line of the road, state it so, and give the width of the road*].

—— ——,
—— ——,
—— ——,
Referees.

☞ These forms can easily be varied by referees to meet every case. Great care should always be taken to make descriptions of roads and alterations full and accurate, and to have the different points fixed in them so plainly and accurately set forth that no one can be misled.

No. 47.

Notice to remove Fences.

To Mr.

Whereas, the undersigned, commissioners of highways of the Town of , in the County of , have laid out a public highway, by an order, dated 18—, and duly filed with the town clerk of the said town, which said road passes through enclosed lands owned [*or occupied*] by you; and whereas our determination in the matter of laying out such road has not been appealed from: Now, therefore, please to take notice, that you are required to remove your fences from within the bounds of said highway within sixty days after service hereof.

Yours, &c.,

A. B.,
C. D.,
E. F.,
Commissioners of Highways.

If the determination of the commissioners has been appealed from, make the notice accordingly.

No. 48.

Article 5, Section 111.

Commissioners' Order to remove Encroachment.

Whereas, a highway was laid out in the Town of , in the county of , on the day of 1859, by the commissioners of highways of the said town [*or by* , *and* , *three of the judges, &c., or referees, &c., as the fact is*], beginning [*insert description as in the order, including a statement of the width of the road*], which road is encroached upon by the fence of , to the extent of [*state how much*] on the *north* side [*describe where*].

Now, therefore, we, the commissioners of highways of the said town [*state the attendance of, or notice to, all the commissioners, unless they all sign the order*], do hereby order that such fence be removed, so that such highway may be of the breadth originally intended.

N. R.,
P. Y.,
D. M.,
Commissioners of Highways.

No. 49.

Article 5, Section 111.

Notice to remove Encroachment.

To Mr.

Sir: Please to take notice, that an order, of which the annexed is a copy, has been duly made by the commissioners of highways of theTown of , in the County of , and that you are hereby required to remove the fence therein specified, as encroaching upon the highway, within sixty days after service hereof.

N. R.,
P. Y.,
D. M.,
Commissioners of Highways.

July 16, 1859.

No. 50.

Summons to Freeholders in case of Encroachment.

ONEIDA COUNTY, *ss:*

To any constable of the Town of Camden, in said county:

You are hereby commanded to summon twelve freeholders of said town, to meet on the 20*th day of June instant*, at the *dwelling house of John Jones*, in said town, to inquire into the matter of an alleged encroachment upon the highway in said town leading from [*here describe the place and the alleged encroachment, as in the order*]; and you are to give at least three days' notice to the commissioners of highways of said town, and to *Abraham Field*, of the time and place at which such freeholders are to meet.

—— ——,

Justice of the Peace.

Oath of Jury.

You, and each of you, do swear, in the presence of Almighty God, that you will well and truly inquire whether any such encroachment as is alleged has been made, and by whom, on the highway now in question.

No. 51.

Certificate of Jury.

COUNTY, *ss:*

The undersigned, freeholders of the Town of Camden, in said county, having met on the day of the date hereof, at the *dwelling house of John Jones*, in said town, pursuant to a summons from *William Walker*, Esq., a justice of the peace of the said county, to inquire into the matter of an alleged encroachment on the highway, in said town, specified in an order of the commissioners of highways of said town, dated , 18 , a copy whereof is hereto annexed, having been duly sworn, and having heard the proofs and allegations which were submitted, do certify that an encroachment on the said highway has been made, and that the same was made by *John Jones*, the present occupant [*or by Samuel Smith, a former occupant*].

And the particulars of such encroachment are as follows, viz.: [*here describe the encroachment, with certainty as to its location and extent.*]

To be signed by the freeholders.

, *June* 20, 18

No. 52.

Warrant for Collection of Costs.

ONEIDA COUNTY, *ss:*

To any constable of the Town of Camdenin said county:

You are hereby commanded to levy of the goods and chattels of *John Jones*, of the said town, the sum of *four dollars and eighty-eight cents*, for the costs of an inquiry into the matter of an encroachment on the highway in said town [*specify the place*], which said encroachment was certified to have been made by the said *John Jones* [*or by Samuel Smith, a former occupant of the land where such encroachment is made, which is now occupied by said John Jones*], by twelve freeholders of the said town on the day of , 18 , the costs of which inquiry are settled and adjudged by me at the sum above mentioned, and bring the said sum of money before me without delay.

——— ———,
Justice of the Peace.

July 1, 1859,

No. 53.

Certificate of Jury when no Encroachment is found.

The same form as No. 51, to the word certify, and then add,

That no encroachment has been made, as was alleged, and we also ascertain and certify the damages which *John Jones*, the present occupant, has sustained by these proceedings, at *five dollars and forty-four cents.*

To be signed by the freeholders.

, *June* 20, 18

INDEX

TO

FORMS AND INSTRUCTIONS.

INDEX.

A

B

BRIDGES, ERECTION OF, REPAIRS, PRESERVATION, &c.

BRIDGE COMPANIES.

C

COMMISSIONERS OF HIGHWAYS.

D

E

F

FERRIES, REGULATIONS OF.

FERRY COMPANIES, FORMATION OF.

G

H

HIGHWAYS, BRIDGES, &c., GENERAL LAW PERTAINING TO.

HIGHWAYS, AND ENCROACHMENTS THEREON.

I

J

L

LAW RELATING TO ELECTION, QUALIFICATIONS AND DUTIES OF COMMISSIONERS OF HIGHWAYS.

M

MISCELLANEOUS PROVISIONS OF A GENERAL NATURE.

N

O

P

PLANK AND TURNPIKE ROADS.

R

ROADS, LAYING OUT OF, &c.

S

T

V

W